IS OUR WATER
SAFE TO DRINK ?

IS OUR WATER SAFE TO DRINK ?

A GUIDE TO DRINKING WATER
HAZARDS AND HEALTH RISKS

J. GORDON MILLICHAP
M.D., F.R.C.P.

*Professor Emeritus of Pediatrics and Neurology,
Northwestern University Medical School,
Children's Memorial Hospital,
Chicago, Illinois*

*Formerly, Pediatric Neurologist, Mayo Clinic,
Rochester, Minnesota*

PNB • Publishers
Chicago

Published and Distributed Throughout the World by

PNB • Publishers

P.O. Box 11391
Chicago, Illinois 60611, U.S.A.

Printed in the U.S.A.
Reprinted, 1995, 1996

ISBN 0-9629115-5-0
Library of Congress Catalog Card Number: 95-67490

1. Water microbiology. 2. Water pollutants, Chemical -
toxicity. 3. Water supply - analysis, treatment.

To my dear wife, Nancy,
who provided invaluable advice and suggestions,
and
To my niece and legal consultant, Kathleen Haviland Esq.,
who kept me informed of most recent governmental
publications.

PREFACE

The safety of our drinking water is often taken for granted in America. In recent years, however, environmentalists and the media have drawn attention to the dangers of ground water pollution and the health risks of lead, chlorine, pesticides, organic chemicals, and various microorganisms that have been found to contaminate our public water supplies. Outbreaks of waterborne diseases are a common occurrence and have involved entire city populations, sometimes leading to serious complications and even fatalities. The potential carcinogenic effects of long-term exposure to certain organic chemicals in our water supplies are under governmental scrutiny.

"Is our water safe to drink?" is a frequent question for the Environmental Protection Agency (EPA) posed by members of congress who are in the process of overhauling the 1974 Safe Drinking Water Act. Do we need to spend more on improved water purification technologies and a tightening of EPA standards for potential cancer-causing chemicals, or should we cut costs and limit regulation of contaminants to those known to cause the greatest health risks, e.g. lead. The former solution would result in criticism by state and local governments who find the present regulations to be costly and sometimes impractical, particularly for smaller municipalities, while the latter would cause an outcry from environmentalists who are pressing for more stringent controls.

Public concern with both the quality and safety of drinking water has become so compelling that many consumers now rely almost exclusively on bottled water for drinking purposes. To learn that regulations governing the purity and safety of bottled water are often less stringent than those for municipal water supplies may come as a surprise to many. The education of consumers regarding the sources of water contamination, treatment processes, the recognition of symptoms of waterborne diseases, and home methods for prevention and control of drinking water hazards should limit the spread of outbreaks of diarrheal illness and minimize the dangers of chronic exposure to lead and other chemical pollutants.

In twelve chapters, this book provides a guide to drinking water hazards and health risks, their sources, recognition, prevention, and their control. In addition to the deficiencies and hazards of public water systems, those of well water and bottled water are also covered. The chief consumer concerns about drinking water quality and safety are outlined, and water treatment units for the purification of water in the home are described. Consumers who are knowledgeable about the causes, symptoms, and signs of waterborne diseases and intoxications may avoid some of the hazards and health risks associated with contaminated drinking water.

J. Gordon Millichap, M.D., F.R.C.P.

TABLE OF CONTENTS

IS OUR WATER SAFE TO DRINK ?

A GUIDE TO DRINKING WATER HAZARDS AND HEALTH RISKS

CHAPTER 1

DRINKING WATER SAFETY
AND HEALTH RISKS

We expect our drinking water to be safe. When we turn on the faucet, fill up a glass, and drink, we should not have to think about water contaminants and health risks. In recent years, however, public concern with both the quality and safety of drinking water has heightened, largely because of reports in the media of ground water pollution with pesticides, other organic chemicals, chlorine, sodium, lead, radon, and various bacteria. Before filling our glass or coffee pot with water in the early morning, we now allow the faucet to run until the water is cold, hoping to avoid the excess levels of lead that have leached from old lead pipes or from lead solder in newer homes.

The potential health risks of waterborne infections and chemical contaminants are so compelling that many consumers rely almost exclusively on bottled water for drinking purposes. Bottled water is now preferred by millions of Americans, and the cost is enormous. Consumers may be surprised to learn that regulations governing the purity and safety of bottled water are often less stringent than those for municipal water supplies.

Microorganisms (bacteria, viruses, parasites) are the major causes of waterborne illnesses. An estimated 25,000 deaths per day were attributed to consumption of contaminated water in the year 1980 worldwide, and 25 percent of hospital admissions were for illnesses related to polluted water. Waterborne gastrointestinal infections account for 80 percent of all diseases (WHO). These are international statistics, and third world countries account for the major proportion of the population affected.

In the United States and other industrialized countries, the incidence of disease caused by waterborne microorganisms is relatively low, but outbreaks are sufficiently frequent to indicate deficiencies in public water system technologies. Gastroenteritis linked to contaminated water supplies is particularly frequent and hazardous in the young, elderly, or in patients with AIDS and malignant disease that compromise immune systems.

Pathogenic, disease-causing microorganisms are present in untreated surface and ground water used as raw sources for drinking water reservoirs. Without efficient filtration and disinfection to limit or eliminate these pathogens, our health can be seriously endangered, irrespective of age, nutritional status, and freedom from chronic disease. Drinking water that is untreated, inadequately treated, or contaminated after treatment is the cause of most waterborne disease.

DRINKING WATER RESOURCES

The two main sources of raw water are surface waters and ground waters. *Surface waters* include rivers, streams, lakes, and reservoirs fed by rainfall precipitation and run-off. *Ground waters* are deep and shallow underground reservoirs called "aquifiers," rivers, wells and springs, fed by the slow percolation of surface waters through soil and subsoil layers. Aquifiers lie below the water table - the level at which water is at atmospheric pressure. They are water bearing formations of sand, gravel, limestone or other porous material. Those situated below a confining layer of impervious rock will yield water under pressure, and are called artesian wells, named after the province Artois in France.

Surface raw waters are replaced relatively rapidly, whereas ground water replacement is slow, sometimes in the thousands of years. These differences in replacement rates are reflected in the varying speeds of purification after contamination. Surface waters are open to the atmosphere and are vulnerable to microbial contaminated by sewage and animal fecal matter in run-off. They lose contaminants rapidly once the source of pollution is removed. In contrast, ground waters in deep aquifiers are usually more protected than surface resources. Because of years of slow percolation and filtration, ground water is generally free of turbidy and pathogenic microorganisms and is the preferred source of potable water. When it does become contaminated with inorganic or organic chemicals it may stay contaminated, because the water is virtually stagnant and the source of pollution is often difficult to remove.

Dowsers and the Divining Rod. The divining rod, a sixteenth century device, continues to be used in search for water in subterranean springs in areas where water is scarce and difficult to locate. The success of the dowser and his rod is controversial. Scientific studies are reported to show that the Y-shaped forked stick, metal rod or wire will twist and dip in the dowser's hands when the squeezing force is relaxed inadvertently, in response to fatigue, without regard for environmental factors. The finding of hidden water is considered coincidence and by chance. Nonetheless, the dowser has his following in some geographic areas.

SOURCES OF WATER POLLUTION

Most raw surface waters and many ground water sources are contaminated by bacteria, viruses, and other organisms, even in the United States. Some of these are pathogens and capable of causing disease. In a 35 years study of waterborne disease outbreaks caused by deficiencies in public water systems, more than two thirds were due to inadequately treated ground and surface water. A bacterial pathogen was identified as the likely cause in 50 percent of outbreaks and chemical pollutants in only 4 percent. Untreated or less frequently disinfected ground waters were implicated as bacterial sources more often than surface waters.

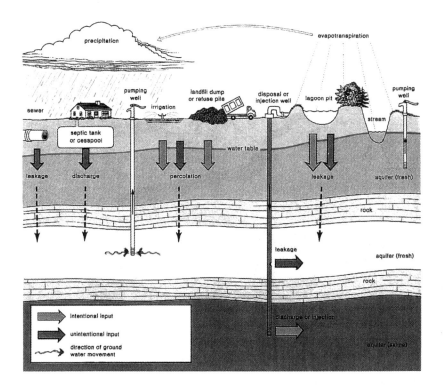

Figure 1-1. Sources of ground water pollution by discharge, percolation, and leakage from septic tank, irrigation, landfill, lagoon pit, or polluted stream.

From U.S. Environmental Protection Agency, Office of Water Supply and Solid Waste Managemnt Programs, "Waste Disposal Practices and Their Effects on Ground Water." (Washington, D.C., U.S. Government Printing Office, 1977).

The number of outbreaks attributed to parasites and viruses has increased as technologies for their identification have been developed and improved. *Giardia lamblia* occurs widely in the United States and is the most commonly identified organism associated with waterborne disease outbreaks. Parasites such as *Cryptosporidium* and *Blastocystis hominis* have been recognized as pathogens more recently, particularly in patients with compromised immune systems, and these organisms are an emerging cause for concern. Due to the small size of these organisms and their resistance to disinfection, they sometimes survive conventional methods of water treatment. More stringent methods designed to remove these disease causing parasites are being installed in response to EPA requirements. The list of potential waterborne disease organisms will probably continue to grow, as technologies for their identification become more sophisticated and more generally available.

In addition to microorganisms, water contaminants include chemicals, inorganic and organic, and radionuclides. Inorganic chemicals are metals and salts, not containing carbon, many of which are naturally occurring, such as arsenic, fluoride, and nitrate. Chemicals such as lead enter water as a result of leaching from lead pipe and lead-based solder pipe joints, either in the distribution system or in the home plumbing. Industrial waste and pesticide and fertilizer use are other sources of inorganic chemical contamination.

Organic chemicals are either natural or synthetic compounds and chlorine containing disinfection by-products, all containing carbon. There are four main categories: 1) Synthetic organic chemicals (SOCs); 2) Volatile organic chemicals (VOCs); 3) Polychlorinated biphenyls (PCBs); and 4) Trihalomethanes - disinfection by-products.

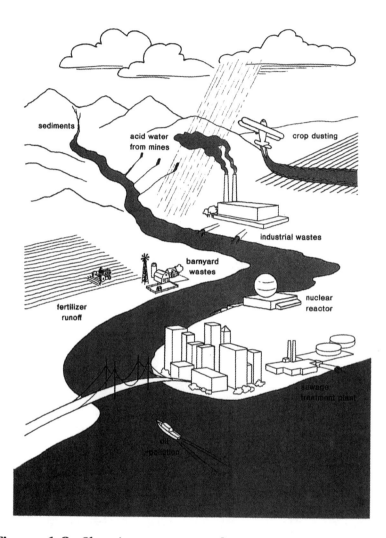

Figure 1-2. Showing sources of water pollution from industrial wastes, fertilizer runoff, crop dusting, nuclear reactor, and sewage treatment plant.

Reproduced and adapted from U.S. Environmental Protection Agency, Office of Water Supply and Solid Waste Management Programs, "Waste Disposal Practices and Their Effects on Ground Water." Executive Summary. (Washington, D.C., U.S. Goverment Printing Office, 1977).

The carcinogenic properties of organic chemicals as well as radionuclides are a main cause for concern. Public health authorities have in the past relied more on filtration and disinfection treatment processes rather than water monitoring to ensure the safety of drinking water, because of the complexity of laboratory analyses. In recent years, more communities are using treatment processes other than disinfection to remove inorganic and organic contaminants. The cost of meeting stricter water quality standards is particularly high for smaller communities.

In the State of Illinois, beginning January 1993, all supplies are required to monitor for 29 synthetic organic pesticides and herbicides. All surface water supplies must test quarterly until no SOCs are detected or the level of contaminants is below the maximum contaminant level (MCL). Ground water supplies will test for most of the SOCs before December 31, 1995.

WATER PURIFICATION TECHNOLOGIES

Water treatment processes provide protection barriers between the consumer and waterborne pathogens and disease. The kinds of drinking water treatments in the United States vary and their complexity depends on the size of the community served, the purity of the water source, and other factors. Several different processes are commonly employed, and the combination of processes is called a *process train.*

Conventional Treatment. The most common process train for surface waters consists of disinfection, coagulation, flocculation, sedimentation, and filtration, followed by a second process of disinfection (Table 1-1). Additional steps frequently employed include preoxidation, preaeration, adsorption, and presedimentation.

At the simplest level, raw water is supplied to the consumer untreated. This practice applies usually to water from ground water reservoirs. The first level of treatment is disinfection, generally with chlorine or chloramines, and sometimes ozonation. Utilities with protected surface waters use the disinfection process.

TABLE 1-1. Water treatment process trains.

Purification Process	Treatment Steps
1. Chlorination	Addition of *chlorine*
2. Filtration	Chlorination followed by *Filtration*
3. In-line Filtration	Chlorination, addition of *Coagulant*, Filtration
4. Direct Filtration	Chlorination, Coagulant, *Flocculation*, Filtration
5. Conventional Treatment	Chlorination, Coagulant, Flocculation, *Sedimentation*, Filtration, *Chlorination*

The second level or process train is chlorination followed by filtration, as shown in Table 1-1. Filtration through sand or coal removes particles and reduces turbidity. At the next level, *in-line filtration*, a coagulant is added before filtration which alters the state of suspended solids and facilitates their removal by filtration. Finally, in the process of *direct filtration,* a flocculation step is added before filtration which improves the efficiency of particle removal; and a disinfectant is applied at the end, as well as the beginning of the process train.

The estimated reductions in waterborne diseases following treatment of water supplies worldwide are 80 to 90 percent for the bacterial diseases, cholera, typhoid, and leptospirosis; 40 to 50 percent for parasitic diseases, ascariasis and amoebic dysentery; and 10 percent for viral hepatitis and enteroviruses. Treatment process trains are much more effective in the prevention of waterborne diseases due to bacteria than those caused by parasites and viruses. Legionella bacteria are primarily a contaminant of distribution systems rather than water sources, and the number of cases of Legionaires disease attributable directly to drinking water is unknown. Legionella should be removed or inactivated by filtration and disinfection in source waters.

In geographic areas with hard waters obtained from limestone formations, softening agents are added which precipitate the calcium and magnesium. Several other specialized techniques are also used to remove iron, manganese, and sulfide, and to correct the taste, odor, and staining problems. Volatile organic solvents such as trichlorethylene and tetrachloroethylene, found in ground water, private wells and hazardous waste sites, are removed by adsorption on granular-activated carbon or by aeration. Other treatment processes used frequently include fluoride addition, and corrosion control, and those employed sparingly are activated alumina, ion exchange, and reverse osmosis.

The costs of these water purification technologies vary according to the size of the plant, the contaminants requiring removal, and type of disinfectant employed. The cost per capita in small communities is greater than in large systems, and some treatment techniques are not available to systems with limited budgets. Practical alternatives are needed when small communities are supplied with raw water requiring more efficient purification.

HEALTH RISKS OF WATER TREATMENTS

The Safe Drinking Water Act Amendments of 1986 required the Environmental Protection Agency (EPA) to evaluate the health risks and benefits of water treatment chemicals and their by-products. Congress was concerned that these technologies may be trading one health risk for another. Outbreaks of acute enteric, or intestinal, illness due to pathogenic organisms in untreated water may be lessened while the incidence of cancer and of other long-term disabilities caused by chlorine disinfection and other chemicals in treated water might be increased.

Chlorination Health Risks. Compared to the health risks of waterborne diseases caused by parasites and viruses, the potential risks of chlorine by-products are less well documented and the long-term toxic effects are poorly understood. Studies in humans are conducted in adults, and testing for adverse effects in children and the unborn fetus, at the most vulnerable age periods, are entirely lacking.

In order to eliminate the formation of

trihalomethanes (THM), the most toxic of these chlorine by-products, some water purification plants have substituted alternative disinfectants to chlorine, including chloramines, chlorine dioxide, and ozone. For smaller surface water systems, however, the increased cost of these alternatives may be impractical. The continued use of chlorine is more economical but the health risks are probably magnified.

Coagulant Residua Health Risks. The use of aluminum and iron salts, sulfates, and polymers in the purification of water may introduce hazards in some individuals, particularly when coagulants are present in high concentration. *Aluminum* in treated surface water varies widely, and levels higher than 0.2 mg/L cause discoloration. Water with higher levels may induce encephalopathy and dementia in patients with kidney disease undergoing dialysis. Aluminum has been linked with Alzheimer's disease.

Ferric iron can cause staining of laundry and other esthetic changes, including taste impairment, if the concentration is at or greater than the EPA maximum contaminant level of 0.3 mg/L. Levels of iron in treated water are generally within a safe range, but if the concentration exceeds the MCL and the blood ferritin is elevated to twice the normal level, the risk of heart attack in males is doubled.

Sulfates and activated silica coagulants are considered nontoxic, but synthetic polymers containing the monomer, acrylamide, are potentially toxic and subject to government regulation.

Corrosion and pH Health Risks. The health risks associated with corrosion and pH control are related to 1) the release of lead from plumbing and, to a lesser degree, cadmium, zinc, copper, iron, and asbestos; and 2) the addition of sodium containing salts to promote alkalinity and increase carbonate ions which lessen corrosion but add to the risks of high blood pressure in susceptible individuals.

When natural carbonate ion is present in raw water, the adjustment of pH toward alkalinity of 8 to 10 by adding sodium hydroxide is sufficient to reduce corrosion of pipes. If carbonate ion is lacking, the addition of lime is required to provide a protective coating of relatively insoluble carbonate salts.

Corrosion inhibitors such as sodium phosphate and silicate are also employed in some distribution systems. These substances react with the metal lining of the pipes to form a protective coating. The leaching of lead from lead pipes in older homes and from lead-based solder on copper pipes and pipe-joints in newer housing is lessened by these corrosion inhibitors.

Lead enters drinking water primarily by corrosion of interior plumbing, faucets, service lines, goosenecks, and solder. Corrosion is of greatest concern in distribution systems with soft, acidic waters, of pH less than 6.5. The health effects of exposure to lead are discussed in detail in Chapter 5, and other heavy metal contaminants are included in Chapter 6.

Asbestos occurs in drinking water as a contaminant at the source or as a result of corrosion in the distribution system. The EPA acknowledges equivocal evidence of a risk of cancer from asbestos-cement pipe and proposed a Maximum Contaminant Level Goal (MCLG) of asbestos fibers in drinking water standards.

Post-Precipitation, the accumulation of insoluble

salts in water after the treatment process, can cause clogging of filters, pipes, and faucets. The precipitation of calcium carbonate and aluminum hydroxide results in high turbidity, a problem that can be prevented by adjusting the pH.

Post-precipitation of manganese dioxide results in black residues in water, and is controlled by oxidation with a secondary disinfectant applied after filtration. If ozone is used, there is no residual chemical, but with chlorine oxidants, health risks may be increased. Post-precipitation is also prevented by adding a complexing agent, sodium hexametaphosphate.

Ion Exchange Health Risks. While ion exchange resins remove toxic metals and other ionic contaminants from drinking-water, they add sodium and aluminum. Patients with high blood pressure treated with low-sodium diets may suffer ill effects.

Lime softening of raw water, a chemical process involving the addition of lime (calcium hydroxide) or lime and soda ash (sodium carbonate) to precipitate calcium and magnesium, also adds sodium and aluminum as unwanted contaminants with attendant health risks.

Reverse osmosis is another water cleaning treatment that uses cellulose acetate membranes and high pressures to remove inorganic (salt) and organic compounds (eg. trihalomethane). This process also involves the addition of various chemicals, including the complexing agent sodium hexameta-phosphate, that prevent precipitation of essential nutrients, calcium carbonate and magnesium salts. Apart from sodium and fluoride, the health risks of additives involved in these water treatment technologies have not been studied in detail.

Activated Carbon Health Risks. Activated carbon is used to remove organic compounds and improve the esthetic qualities of drinking water, especially taste and odor. Unfortunately, granular activated charcoal is an ideal environment for the growth of bacteria because it interferes with chlorination. It also introduces potentially toxic organic compounds such as polychlorinated biphenyls, by polymerization and chlorination of phenols on the surface of the carbon.

This method of water purification also removes radon and other radionuclides. The adsorption of these hazardous compounds complicates the land disposal of the carbon.

Few data are available on the efficiency of activated carbon adsorption and the health effects of the contaminant chemicals identified. More research is needed to establish the potential health risks of microorganisms and organic chemicals generated by this method of water treatment.

Air Stripping Health Risks. Air stripping, also called aeration, removes volatile organic compounds (VOCs) such as trichlorethylene and tetrachloroethylene from water by transferring them to the air. It does not remove non-volatile organic chemicals, of equal or greater concern than VOCs.

Radon is also removed from ground water by air stripping and may pose a risk to treatment plant workers inhaling contaminated air in the proximity of the aeration system. The inhalation of radon is more toxic than ingestion and the risk of radon-related lung cancer may be increased. The EPA determined that exposure to natural radon volatilized from drinking water could result in an increase of 600 lung cancer deaths per year in the United States. Studies are in progress to evaluate the significance of these potential transfer risks.

DRINKING WATER CONTAMINANTS AND DISEASE OUTBREAKS

The safety and quality of drinking water in the United States are generally enhanced by the water treatment processes commonly used in most public water systems. The removal of microorganisms by disinfection and filtration has dramatically reduced the incidence of waterborne diseases such as typhoid fever, cholera, and hepatitis, and the banning of lead in pipes and solder has reduced the exposure of millions of Americans to lead and the risks of lead poisoning.

The Safe Drinking Water Act of 1974 and the Amendments of 1986 set standards of drinking water quality to protect our health, but deficiencies in the regulatory program enforcing these goals and the health risks of inadequate treatment processes and their chlorination by-products are of concern. Pathogenic microorganisms, and particularly *Cryptosporidium*, are the continuing major cause of waterborne disease outbreaks in the United States.

Of a total of 96 outbreaks of waterborne disease affecting 46,712 confirmed cases reported to the Center for Disease Control and Environmental Protection Agency in the six year period 1986-1992, only three were traced to *Cryptosporidium*. However, a total of 16,551 persons were affected, accounting for 35 percent

TABLE 1-2. Infectious Causes of Waterborne Disease Outbreaks Attributed to Contaminated Drinking Water and reported to the CDC and EPA, 1986-1992.

Organism	Outbreaks		Cases		States
	No.	Percent	No.	Percent	No.
AGIUE[1]	58(60%)		18,603	(40%)	22
Giardia	20(21%)		1,681	(3.6%)	10
Cryptosporidium	3 (3%)		16,551	(35%)	3
Norwalk-like	5 (5%)		6,415	(13.7%)	7
Shigella Sonnei	4 (4%)		2,875	(6%)	4
Salmonella	2 (2%)		70	(0.1%)	2
Campylobacter	1 (1%)		250	(0.5%)	1
E.coli	1 (1%)		243	(0.5%)	1
Cyanobacteria	1 (1%)		21	(0.04%)	1
Hepatitis A	1 (1%)		3	(.006%)	1

[1]Acute Gastrointestinal Illness of Unknown Etiology.
Derived from "Morbidity and Mortality Weekly Reports"
CDC and EPA, 1990-93.

of all cases of waterborne infections reported. In the spring of 1993, after the completion of this report, a massive outbreak of cryptosporidium infection affecting more than 400,000 of the population of Milwaukee was attributed to an inadequate public water filtration system. Despite unprecedented increases in turbidity of treated water at one of the city's plants, the cause of the problem was undetected and the contaminated water plant remained open for almost three weeks.

Since the EPA regulation of chlorination by-products, especially trihalomethanes (THMs), in 1979, many public water systems have tried to minimize the health risks of these carcinogenic chemicals by using alternative disinfectants to chlorine. The cost effectiveness and health risks of the new methods of disinfection require further study.

Conventional drinking-water treatment technologies are not entirely satisfactory for the removal of industrial solvents, agricultural pesticides, and radioactive substances, including radon. Smaller community systems find the newer and more efficient methods too costly and impractical.

Lead pipe and solder corrosion is a significant source of lead exposure and poisoning, particularly in older homes and in communities supplied with soft, highly corrosive water. The 1993 change in the EPA water lead level standard from 50 parts per billion (ppb) down to 15 ppb should reduce the exposure to lead for millions of Americans. Corrosion control is important in the lowering of lead concentrations and other metal by-products in our drinking water as well as its effect on the esthetic quality of water.

DRINGING WATER ACT OVERHAUL

The 1974 Safe Drinking Water Act, last amended in 1987, expired in 1991 and has remained viable by annual appropriations. A Safe Drinking Water Reauthorization Bill passed the Senate in May 1994 but failed to pass the House of Representatives.

This overhaul of the 1974 law would have dealt with funding, water contaminants, small water systems, monitoring, and health standards. A new state revolving loan fund would make money available to local goverments to construct and repair outdated water systems. The regulation of contaminants by the EPA would be limited to those posing the greatest health risks. Small water system requirements and regulations would be waived or made more affordable. Monitoring of contaminants would be reduced, and health standards for drinking water supplies would be eased. In developing standards to limit potential cancer-causing agents, the EPA was required to weigh health risks and the costs of risk reduction. The zero-risk Delaney Amendment would no longer apply to carcinogenic contaminants in water.

The bill was praised by state and local governments but criticized by environmentalists. The The Senate had voted to loosen the standards for drinking water after having tightened them in 1987. With the current public concern regarding environmental factors in the cause of cancer, any legislation designed to limit standards and monitoring of cancer-causing chemicals was likely to fail. The recent outbreak of *Cryptosporidium,* a waterborne parasite affecting nearly half a million residents of Milwaukee and contributing to 100 fatalities, was reason for stricter standards and closer monitoring of our public water supplies.

Expenditures for current regulations allot 75 percent of costs for 1) filtering and disinfecting water to remove bacteria and viruses; and 2) reducing lead levels in drinking water. Despite these appropriations, more than 46,000 cases of waterborne infections and 96 outbreaks of acute diarrhea were reported between 1986 and 1992. Three quarters of these cases were attributed to an "acute gastrointestinal illness of unknown etiology" or to cryptosporidium infection. The number of cases was increased ten fold by the Milwaukee outbreak in 1993.

WATER TREATMENT AND MEDICAL DIAGNOSTIC INADEQUACIES

Sporadic cases of waterborne disease may occur as a result of an overwhelming contamination or a transient deficiency in the public water purification technologies. Widespread outbreaks of disease result not only from a fault in the drinking water treatment systems, but also because of a failure of medical and public health diagnostic services.

The massive outbreak of cryptosporidium infection in Milwaukee was a reflection of 1) an inadequate filtration system and inefficient water-quality monitoring in a large city's public water supply; and 2) a delay in diagnosis in patients who sought medical help. The slow response of physicians to recognize the cause of the parasitic waterborne illness was due to 1) a misdiagnosis of viral gastroenteritis or "intestinal flu," without further investigation; and 2) failure to request and use special testing procedures required for the detection of *Cryptosporidium* in stools examined for ova and parasites.

In addition to the shortcomings of the municipal water system and medical services, the private citizen was also culpable for failing to recognize the potential hazard of water turned turbid at the faucet. A sudden change in appearance of the water should have alerted the consumer to a probable contaminant and the need to use alternative drinking water sources, such as bottled water.

Blastocystis hominis is another parasite emerging as an important cause for concern in some patients with acute gastrointestinal illness. The failure of physicians to recognize this organism as a pathogen, especially in immune-compromised patients, can lead to chronic incapacitating diarrhea and severe malnutrition. Many of the cases reported as "acute gastrointestinal illness of unknown etiology" may be attributed to these under-recognized parasitic infections.

PREVENTION AND CONTROL OF WATERBORNE DISEASES

The safety of our drinking water and the prevention of waterborne outbreaks of infection will depend on:

1) Improvements in the methods of fitration and disinfection as well as monitoring of the water supply;

2) Education and cooperation of the medical community to recognize and expedite the diagnosis of these pathogens in their patients; and

3) Education of the public to be alert to signs of contamination of their drinking water and the need to take immediate precautions to avoid health risks.

Since 60 percent of outbreaks and 40 percent of cases of waterborne infections are of unknown cause, Federal and State funds should be made available to investigate the origin of these infections and determine more efficient methods for their diagnosis and prevention at the source.

How   can   government   regulate
contaminants   that   "pose   real   health   risks"   if
we   are   unable   to   recognize   the   organisms   and
chemicals   responsible   for   a   majority   of
waterborne   diseases?

MacKenzie WR, Hoxie NJ, Proctor ME, et al. A massive outbreak
    in Milwaukee of Cryptosporidium infection transmitted
    through the public water supply. N Engl J Med July 21
    1994;331:161-7.

Mandell GL, Douglas Rg Jr, Bennett JE. (Eds). Principles and
    Practice of Infectious Diseases. 3rd edition. New York,
    Churchill Livingstone, 1990.

Millichap JG. Environmental Poisons in Our Food. Chicago, PNB
    Publishers, 1993.

Olson ED. Think before you drink. The failure of the Nation's
    drinking water system to protect public health.
    Executive Summary. Natural Resources Defense Council.
    September 1993.

Ram NM, Calabrese EJ, Christman RF. Organic Carcinogens in
    Drinking Water. Detection, Treatment, and Risk
    Assessment. New York, John Wiley & Sons, 1986.

U.S. Center for Disease Control and EPA. Waterborne disease
    outbreaks, 1986-1988. Morbidity and Mortality Weekly
    Report. March 1990;39:SS-1.

Idem. 1989-1990. Ibidem. 1991;40:SS-3.

Idem. Surveillance for waterborne disease outbreaks - United
    States, 1991-1992. Ibidem. Nov 1993;42:SS-5.

U.S. Congress House Floor Brief. Safe Drinking Water Act of
    1994. Oct 7, 1994.

U.S. Environmental Protection Agency. Comparative Health
    Effects Assessment of Drinking Water Treatment

Technologies. Washington, D.C., Office of Drinking Water, 1988.

Idem. Drinking Water Protection. A general overview of Safe Drinking Water Act Reauthorization. Office of Water. February 1994.

Idem. Administration's recommendations for Safe Drinking Water Act Reauthorization. Office of Water. Feb 1994.

Idem. Drinking Water Issue. Risk reduction and drinking water standards. State capacity. Small system compliance. Source water protection. Office of Water. February 1994.

Idem. Myths and Facts. The "Pineapple Pesticide." Standards.Office of Water. February 1994.

Idem. Weekly Water Notes. Special Edition. House drinking water agreement. EPA's Office of Water, Aug 23, 1994.

U.S. General Accounting Office. Drinking Water. Stronger efforts essential for small communities to comply with standards. Report to the Chairman, Environment, Energy, and Natural Resources Subcommittee, Committee on Goverment Operations, House of Representatives. March 1994.

World Health Organization. Guidelines for Drinking-Water Quality. Volume 2. Health Criteria and Other Supporting Information. Geneva, 1984.

Idem. Guidelines for Drinking-Water Quality. 2nd Edition. Geneva, Switzerland, Office of Publications, WHO, 1993.

CHAPTER 2

WATERBORNE BACTERIAL DISEASES

I nfectious diarrhea related to waterborne bacteria is a major cause of illness, especially among infants and children. Adults with chronic, debilitating disorders and compromised immune systems also have an increased susceptibility to water-related infections. In developing nations, the incidence of waterborne bacterial gastroenteritis is very high. Approximately 750 million cases worldwide and 5 million deaths annually result from diarrhea. Contaminated drinking water from improper sewage disposal is a major source of infection.

Shigella is the most common bacterial cause of outbreaks of waterborne disease reported to the Centers for Disease Control, Atlanta, U.S.A. In the periods 1972 - 1985 and 1986 - 1992, *Shigellae* organisms were isolated in 12 % and 4 % of outbreaks, respectively. Salmonella was the cause of 5 % and 2 % of outbreaks during these two report periods. *Escherichia coli* accounted for only one percent of outbreaks.

In comparison with bacterial related outbreaks of waterborne disease, the parasitic disease *Giardia lamblia* is much more common. Giardiasis was diagnosed in 39 % and 21 % of outbreaks reported. With the more recent outbreaks of *Cryptosporidium*, affecting close to half a million people who drank a major U.S. city polluted water in 1993, parasites are emerging as the most common cause of water-related gastointestinal disease.

Table 2-1 provides a list of bacterial-related waterborne diseases and the primary sources of drinking water contamination.

TABLE 2-1. Bacteria, Related Diseases, and Sources of Drinking Water Contamination.

Organism	Disease	Source
Shigella	Bacillary dysentery	Human feces
Salmonella spp.	Salmonellosis	H/A feces[1]
Salmonella typhi	Typhoid fever	Human feces
S. parathyphi	Paratyphoid fever	Human feces
Vibrio cholera	Cholera	Human feces
Escherichia coli	Gastroenteritis	Human feces
Yersinia spp.	Gastroenteritis	H/A feces[1]
Campylobacter jej.	Gastroenteritis	H/A feces[1]
Legionella	Legionellosis	Water

[1]Human/animal feces. Adapted from EPA Report, 1989.

SOURCES OF WATER CONTAMINATION

The extent of the contamination of our drinking water by disease-causing bacteria is poorly defined because of the difficulties and costs of identifying pathogens in the laboratory. Concentrations of bacteria in water are affected by several factors, including the incidence of disease in the animal population close to the water source, human disease carriers, sanitation in the community, and the adequacy of disinfection and filtration treatment of water supplies.

Since currently available techniques for the identification of specific pathogenic organisms are not always readily available, the concentrations of common fecal coliform and other indicator bacteria are used to judge the quality of sanitation and level of water contamination by animal and human wastes. Fecal coliform, streptococci, and *E. coli* inhabit the gastrointestinal tract of all warm-blooded animals, and the presence of these organisms in water is an indication of fecal contamination.

These indicator bacteria and pathogenic microorganisms reach the surface water supplies via improperly treated sewage, human waste, street and storm water run-off, agricultural grazing land, and animal excreta. The fecal contamination of raw surface water is a frequent occurrence, and the potential for water contamination by pathogenic organisms is real. Unless the raw supplies of water are treated to reduce or eradicate the pathogens, the health risks incurred by drinking water will be high.

Fortunately, in the U.S.A. waterborne bacterial

diseases are generally prevented by efficient raw water purification, but occasionally, at periods of increased contamination, water treatment plants may be overwhelmed. At times of heavy rainfall, the run off from agricultural lands may lead to concentrations of raw water bacteria that exceed the acceptable safe levels of fecal coliforms and raise the risk of disease-causing bacteria.

While typhoid and paratyphoid fevers and cholera have largely been eradicated in the U.S.A., outbreaks of shigellosis and salmonellosis caused by contaminated drinking water are still a concern. A knowledge of the symptoms and signs of these waterborne infections will lead to earlier diagnosis, identification of the source, treatment, and prevention of further cases.

SHIGELLOSIS (BACILLARY DYSENTERY)

Shigellosis, or bacillary dysentery, is an acute diarrheal illness caused by *Shigella* bacteria. There are four serogroups of *Shigella* with more than 40 serotypes. *S. sonnei* is the most common cause of bacillary dysentery in the United Sates and Europe, accounting for more than two thirds of the cases reported. *S. flexneri,* the Flexner bacillus, and *S. boydii,* the Boyd bacillus, are uncommon causes. *S. dysenteriae,* a severe form, is endemic in the Far East and is rarely encountered in the United States.

Source and Spread of Infection. Shigellosis is a disease of humans, and is spread by person-to-person contact and via fecal contamination of water and food. Animals are not affected. It is often a disease of poverty or overcrowding, when optimum sanitation is disrupted. Recent outbreaks affected large numbers of passengers on cruise ships.

In the past, bacillary dysentery has been reported frequently in institutionalized populations and under wartime conditions, The influence that bacillary dysentery had on military campaigns may have changed history. In the Napoleonic, Crimean, and American Civil Wars, a heavier toll has been ascribed to bacillary dysentery than to war-related injuries.

Bacillary dysentery is the most highly communicable of the bacterial diarrheas. Most cases are the result of transmission from person-to-person. Wide-spread epidemics also occur by exposure of populations to contaminated water or food, directly via handlers who are suffering from shigellosis or by carriers of the organism.

A low dose of *Shigella* bacteria is sufficient to cause illness in a person infected. Tranfer from person to person by direct contact is common when one case is introduced into a family or a close community such as an institution, hospital, ship, or army barracks. The rate of spread of infection is about 40 percent for young children and 20 percent for all ages, when one case has been identified in a household or group.

Children from 6 months to 10 years are at greatest risk, and particularly those of one to four years of age. Preschool children in day care centers are vulnerable, infection being transmitted to the hands via fecal contamination of lavatory seats. Adults often acquire the illness from their children.

Hand transmission is a frequent method of acquiring bacillary dysentery. Mentally retarded children and adults are particularly susceptible. Fecal organisms such as *E. coli* have been recovered from the fingers of the majority of institutionalized patients, and cultures taken from fingers grew *Shigella* in 10 percent of those with positive stool cultures. Fecal excretion of the infecting organism may persist up to four weeks after an attack of shigellosis, if antibiotic treatment is not prescribed. Long-term carriers of the organism may be important in the transmission and epidemiology of shigellosis in some institutions.

In developing countries, where breast feeding is commonly practiced, infants are resistant to shigellosis, and the disease is rare from birth to six months of age. In industrialized societies, however, where bottle feeding is the rule, newborns may lack the specific immune antibodies. If infection is transmitted to the infant, the diarrhea may be severe.

Wells located close to cesspools and privies and contaminated with human fecal material are a frequent source of waterborne shigellosis in developing countries. Septic tanks that empty into lakes or beaches are sometimes a source of water contamination in small communities. Chlorination generally controls the threat of infection from these sources in the United States, but camp water and private wells are occasionally suspect.

Symptoms, Signs, and Diagnosis. Symptoms of acute diarrhea generally begin within one to seven days of contracting the infection. The first symptoms are usually fever and abdominal cramps followed by profuse watery stools. Infants and young children sometimes have febrile convulsions associated with shigellosis and high fever. A neurotoxin formed by *Shigella* bacteria is postulated as a possible cause of the seizures.

As the infection progresses down the alimentary tract, from the small to the large intestine, the stools become smaller in volume and tinged with blood and mucus. Grossly bloody stools are usually complicated by urgency and pain with defecation, termed *tenesmus.*

Stool culture. This classical onset of bacillary dysentery should prompt a stool culture to make a definitive diagnosis. The organisms are present in large numbers during early stages of the attack, and cultures of stools are generally positive for *Shigella.* In later stages, when the infection has localized in the colon and rectum, cultures are taken from ulcers in the intestinal lining, and feces are incubated in special media to ensure a positive culture.

Antibodies in the serum develop only after clinical recovery from infection, and evaluation of the antibody response is generally of no value in the initial diagnosis of bacillary dysentery. Serologic evaluation of patients is helpful in defining the *Shigella* serotype involved and the extent of an epidemic.

Treatment. The presence of blood or mucus in the stools, or persistent fever, diarrhea, and dehydration are indications to consult a physician. Loss of fluids as a result of diarrhea and vomiting must be corrected by drinking carbonated beverages or tea made with boiled water and sweetened with honey. Milk and other dairy products, fresh fruits and raw vegetables, and solid foods in general should be avoided until the diarrhea stops. Then start with the introduction of cooked rice, apple sauce, bananas, jello, chicken, and beef consomme, avoiding milk and fatty foods until recovery is complete.

Infants should continue breast feeding when possible, and electrolyte solutions are prescribed by the physician as supplements when necessary to maintain fluid and electrolyte balance.

Water intoxication. Excess distilled water or glucose water, without added salts, can lead to water intoxication, a complication to be avoided. Improper dilution of infant formula and feeding large amounts of free water or glucose water, especially at times of diarrhea, are common practices in United States inner cities. These practices are reaching epidemic proportions in recent years, and are sometimes the result of improper medical advice. Water intoxication may cause seizures in young infants. The diagnosis is confirmed by a blood test showing a low serum sodium level, or hyponatremia. Any infant with a seizure not responding to anticonvulsant medicines should have a serum sodium determination.

Antibiotics. The use of antibiotics in treatment of shigellosis is controversial. The infection is generally self-limited, and some authorities reserve antibiotic therapy for only the most severely affected patients who have persistent bloody diarrhea or fever and toxemia. The American Academy of Pediatrics recommends antibiotics for small children with shigellosis before readmission to a daycare center, to avoid transmission of the organism to·others.

While antibiotics will often shorten the duration of the illness and eradicate the organism from the stools, some organisms become antibiotic resistant, leading to infections that are chronic and predisposing to a carrier state. The public health reasons to treat all patients with a positive culture and prevent transmission from person to person may be offset by the development of an antibiotic-resistant strain of *Shigella* and a potential source of prolonged infection.

Antimotility drugs. The profuse, watery diarrhea may flush the bacteria through the intestine and prevent the invasion of the colon by the organism and development of ulceration of the bowel wall. Diarrhea is viewed as a protective mechanism.

If antimotility drugs (eg. Lomotil, Paregoric) are prescribed to lessen bowel movement, cramps and the frequency of stools, the duration of the infection can be prolonged, leading to a toxic dilatation of the colon and other complications. Rarely, with some strains of *Shigella,* a "hemolytic uremic syndrome" may develop, a complication exacerbated by antimotility drugs. Despite their ability to alleviate distressing symptoms, drugs of this nature are best avoided.

Complications. Bacillary dysentery is often severe and incapacitating. It may necessitate admission to hospital, but recovery is usually complete. Without antibiotic treatment, the clinical course lasts about one week, but may be extended for one month. Mortality is a rare occurrence, except in debilitated children with a fulminating form, in elderly patients, and in infections caused by *Shigella dysenteriae.*

Complications may include severe dehydration, febrile seizures, pneumonia, eye and urethral infections, and an arthritic syndrome (Reiter's syndrome). A rare, acute and rapidly fatal toxic encephalopathy, a complication in infants, is called *ekiri syndrome*. The *hemolytic uremic syndrome,* with convulsions, impairment of consciousness and coma, complicates infections that produce a Shiga toxin. *E. coli* O157:H7 is another organism known to produce the hemolytic uremic syndrome.

Prevention. 1) A safe water supply is essential for the control and prevention of outbreaks of shigellosis. This is maintained by an optimum level of sanitation and efficient human sewage disposal and treatment. Chlorination of water is responsible for the decrease in *Shigella* dysentery and other enteric bacterial diarrheas in the United States.

2) The spread of infection may be prevented by attention to good personal hygiene and the education and careful supervision of food handlers.

3) Breeding places of insects that carry disease should be removed. The prompt collection and disposal of garbage is a hallmark of efficient sanitation.

4) Transmission from patient to contacts must be minimized by frequent handwashing, disinfected toilet seats, and refrigeration and adequate cooking of food.

5) Breast feeding of infants, particularly in developing countries and communities with inferior standards of hygiene.

6) Early diagnosis, isolation, and reporting of cases, so that the source of infection can be promptly localized and personal and food handler hygiene can be tightened to limit transmission.

7) Treatment with antibiotics in selected cases, to shorten the period of contagion and reduce the number of virulent organisms and carriers.

8) Educate consumers, parents, and children regarding symptoms of dysentery, methods of transmission from person to person, and ways to avoid infection, especially when travelling in developing countries.

9) In countries where hygiene and sanitation are deficient, travelers should avoid untreated drinking water, ice in drinks, soups, unpasteurized milk and milk products, salads, and raw vegetables. Drink only carbonated, bottled water, soft drinks, without added ice cubes, or beer from the bottle or can, not a glass that may have been washed in contaminated water. Eat only fruit that can be peeled, and other foods that have been cooked and are served hot. Use commercial bottled water with sealed cap for cleaning teeth.

10) An attack of bacillary dysentery confers a degree of immunity and may protect the sufferer from further attacks. Some form of oral vaccine is under investigation for protection of persons in endemic areas or communities.

SALMONELLOSIS

The salmonellae are the most ubiquitous group of microorganisms that cause waterborne or foodborne diarrhea. In humans, salmonella infection occurs in three forms: 1) enteric fevers, typhoid and paratyphoid; 2) acute gastroenteritis, the most common form; and 3) focal infections of bone, osteomyelitis, and abscess formation.

Salmonella typhi, causing typhoid fever, and *S. paratyphi* A and B, the organisms of paratyphoid fever, affect only humans and are not natural pathogens for animals. Typhoid fever is now rare in the United States but is a serious disease, with a fatality of 12 - 25 percent before the advent of antibiotics. A carrier state is common after recovery from the acute bacteremia and fever, and the organism establishes permanent residence in the gallbladder without harm to the host. Surgical removal of the gallbladder is often required to make the patient carrier-free.

Typhoid bacteria are transmitted to foods by the hands of carriers, insects, rodents, and water contaminated with infected human wastes. Shellfish grown in polluted water, and unpasteurized milk are sources of infection. Improved hygiene and sanitation, the proper treatment of human sewage, education of consumers regarding food preparation and refrigeration, and immunization of travelers to endemic areas have drastically reduced the incidence of enteric fevers in the United States.

SALMONELLA GASTROENTERITIS

Symptoms. Salmonella gastroenteritis and food poisoning, manifested by diarrhea, abdominal cramps and fever, are caused by the organisms *Salmonella typhimurium* and *S. enteritidis.* Symptoms begin within 6 to 72 hours of contracting infection. More than 2000 different serotypes of *S. enteritidis* have been isolated, but 10 salmonella serotypes account for 75 percent of clinical outbreaks of salmonella food poisoning. The identification of serotypes is important in epidemiology and in recognition of the common source of infection in outbreaks.

Source of Infection. Salmonella organisms

cause disease in animals and humans. The principal reservoirs for infection are poultry, livestock, reptiles, and pets. Foods of animal origin, including poultry, red meat, eggs, and unpasteurized milk, are the major routes of infection. The drinking of contaminated water was the cause in 70 cases identified in two outbreaks reported in a six year period, 1986-92, in the U.S.A. These accounted for only 0.1% of cases and 2% of outbreaks of waterborne disease attributed to contaminated drinking water. Other methods of transmission include playing with pet turtles, and person-to-person passage by the hand to mouth route.

Salmonella is widely disseminated in the food processing industry, and particularly in egg and poultry production. Poultry growers use feeds containing animal by-products that are infected with salmonellae. The salmonellae multiply in the gastrointestinal tracts of the birds, contaminate the surface of the eggs, and persist in the poultry flesh after evisceration. Eggs should be washed and kept refrigerated to minimize the danger of salmonellosis in the consumer. Cracked eggs are a major hazard and should be used only in dishes requiring heat in cooking.

A major recent outbreak of salmonellosis, affecting persons from several States who had eaten a particular brand of ice cream, was traced to a truck container used to transport the bulk ice cream mix. The truck had previously carried a shipment of raw eggs which was found to be contaminated, and the container had not been adequately cleaned and sterilized between trips.

Incidence. Between the years 1955 and 1985,

the number of cases of salmonellosis reported annually in the United States has increased steadily from 4 per 100,000 population to 27 per 100,000. The highest incidence was in children less than 5 years of age and in adults over 70 years. The peak incidence was during the first year of life. The relatively low infection rate after 10 years suggests the development of an immunity to salmonella in adults, except for the elderly.

Most cases occur singly and sporadically, but outbreaks are reported in households and institutions. Severe and fatal infections are unusual, occurring mainly in infants and in patients with AIDS, malignancies, and other immunocompromising disorders.

Infants remain contagious longer than older children and adults, and precautions with fecal contamination and person-to-person contact should be strictly observed for at least six months or until three stool cultures are negative. Antibiotic therapy prolongs the period of excretion of *Salmonella* in the stools and is discouraged, except in patients with invasive disease and complications.

Prevention. Salmonellosis may be avoided by following certain well-established principles of food selection and preparation, by chlorination of community water supplies, and by observing rules of proper sanitation, hand washing and personal hygiene. Eggs and other animal foods should be cooked thoroughly before eating; raw eggs in salads and sauces should be avoided. Infected persons should be excluded from handling food, the sale of turtles for pets should be prohibited, and outbreaks of proven salmonellosis should be reported to health authorities for investigation of the source.

CAMPYLOBACTER INFECTIONS

Symptoms and Complications. The symptoms of *Campylobacter* infections include diarrhea, abdominal pain, and fever. The stools may be bloody, especially with infection in newborn infants. The abdominal pain may mimic an appendicitis. Recovery usually occurs within 48 hours, and a prolonged or severe illness is uncommon.

Complications include convulsions with fever in small children; arthritis, sometimes associated with eye and urethral infections (Reiter's syndrome); and an acute infectious neuritis with generalized paralysis called Guillain-Barre syndrome. One strain of *Campylobacter, C.pylori,* has been linked to gastritis and peptic ulcer.

Guillain-Barre syndrome and *Campylobacter.* *Campylobacter jejuni* organisms were isolated from stool cultures of 30 percent of 46 patients with Guillain-Barre syndrome, compared to only 1 percent of 503 healthy persons in a study in Japan. This finding was highly significant of a relation between the nerve paralyses and *C. jejuni* infection. Furthermore, serological tests in patients with negative stool cultures showed evidence of recent infection with *Campylobacter* in 5 of 29 patients. Therefore, a total of 41% of patients in this study had developed Guillain-Barre syndrome as a result of, or in association with, this waterborne or foodborne infection.

In contrast to this high correlation in Japan, *C. jejuni* infection has been linked to Guillain-Barre syndrome in 18% of cases in the United States and in 14% in the United Kingdom. Other organisms have sometimes been invoked as the cause of this nerve paralysis syndrome, but *C. jejuni* is the most commonly associated agent.

Source of Infection. Transmission of *C. jejuni* occurs by eating or drinking contaminated food, water, and unpasteurized milk, or by direct contact with excreta of infected poultry, cats and dogs, or infected persons, particularly young children. The incubation period from the time of contact to onset of symptoms is 1 to 7 days. Excretion of organisms during recovery is brief, usually only 2 weeks, or as short as 2 days, if antibiotic therapy is used. A carrier state is rare.

Diagnosis. *C. jejuni* can be cultured from the feces, but special methods are necessary. The physician may need to specify the organism suspected, since some laboratories do not stain for *Campylobacter* routinely.

Treatment. The physician may prescribe erythomycin, which shortens the illness and quickly removes the organism from the stools. For young children attending day care centers, antibiotic treatment reduces the period of isolation and prevents spread of the disease.

Control. To avoid infection with *Campylobacter:*
• Chlorinate water supplies and pasteurize milk.
• Wash hands after preparing chicken, wash cutting boards, and cook poultry thoroughly.
• Exclude infected infants from child care centers.

ESCHERICHIA COLI DIARRHEA

Symptoms and Types. Four main types of

diarrheal illness associated with different strains of *Escherichia coli* infection are recognized:

1) Watery diarrhea, often severe and chronic, in infants and young children in developing countries. It may cause dehydration, loss of weight, and a failure to thrive.

2) Watery diarrhea of infants in developing countries, and a self-limited illness of moderate severity. A major cause of *travelers' diarrhea* in adults.

3) Bloody diarrhea, similar to the bacillary dysentery caused by *Shigella,* and associated with fever, vomiting, abdominal cramps, and tenesmus (straining).

4) Bloody diarrhea complicated by hemolytic-uremic syndrome (HUS), hemorrhagic colitis, and purpuric rash. *E. coli* O157:H7 is the strain of organism underlying this syndrome.

The terms *enteropathogenic E. coli (EPEC), enterotoxigenic (ETEC), enteroinvasive (EIEC), and enterohemorrhagic (EHEC),* respectively, have been applied to the *E. coli* strains linked to these four diarrheal syndromes.

Source of Infection. Most diarrheal illness due to *E. coli* is transmitted from infected patients or carriers, or from contaminated food or water. Waterborne *E. coli* accounted for 1 percent of outbreaks of diarrheal disease attributed to drinking water in the six year reporting period, 1986-92, in the United States. A total of 243 cases were involved in only one State.

Undercooked meat is a more frequent cause of outbreaks, particularly for the bovine related strain, *E. coli* O157:H7. In 37 children who ate contaminated hamburger meat in the Seattle region, 95% had severe hemorrhagic colitis, 51% had heart and kidney disease, and 16% had neurological complications. Seizures occurred in 3, stroke in 3, and 3 became comatose. Three (8%) died. This recent outbreak of *E. coli* diarrhea complicated by HUS is a tragic reflection of the hazards of environmental poisons in our food and water.

The incubation period for most *E. coli* strains is 1 to 6 days.

Diagnosis. Bloody diarrhea should indicate the need for a stool culture and serotyping test for *E. coli* O157:H7. Apart from this strain, laboratories have difficulty in differentiating the pathogenic or disease-causing *E. coli* from the normal stool *E. coli*. Children with blood in the stools and bowel intussusception may have *E. coli* O157:H7 infection as the underlying cause of the acute bowel obstruction.

Treatment. Dehydration and loss of electrolytes are corrected by drinking appropriate fluids. Antimotility drugs should be avoided. Antibiotics may be recommended in some cases, but the complication hemolytic-uremic syndrome associated with *E. coli* O157:H7 is not always prevented.

Prevention. 1) Strict observance of proper hygiene and hand washing helps to limit spread of infection in nurseries and households.

2) Ground beef should be thoroughly cooked. Hamburgers should be eaten "well done," not "rare."

3). Public health authorities should be notified promptly of outbreaks of diarrheal illness, in order to identify, isolate, and eradicate the source.

4) Travelers in developing countries are advised to drink only boiled or carbonated water or canned beverages. They should avoid ice, salads, or fruit that

has not been peeled personally. They should select hot dishes to eat.

Prophylactic antibiotics taken before a trip abroad are discouraged by the American Academy of Pediatrics, because of the risks of drug reactions in children and the development of drug resistance. Adults should consult a physician if preventive therapy is contemplated before travelling abroad.

YERSINIA ENTEROCOLITIS

Yersinia organisms cause several clinical syndromes. The most common are *Y. enterocolitica* and *Y. pseudotuberculosis*.

Symptoms. The colitis syndrome is manifested by diarrhea with fever. Abdominal pain and tenderness sometimes mimics appendicitis. Abscess formation, meningitis, pneumonia, kidney infection, arthritis, and skin rash may occur as complications.

The pseudotuberculosis syndrome consists of fever, rash, and abdominal pain. The acute abdominal pain symptoms result from swollen glands and appendicitis.

Source. Animals, chiefly hogs and rodents, carry the organism. Infection is transmitted to humans via: 1) contaminated uncooked pork, unpasteurized milk, or water; 2) contact with animals; and 3) transfusion with packed red cells. Patients with excessive iron storage are particularly susceptible.

Treatment. Antibiotics are indicated in patients with complications and in those with compromised immune systems.

Prevention. Avoid ingestion of inadequately

cooked pork, contaminated water supplies, or unpasteurized milk.

LEGIONNAIRES' DISEASE

Legionnaires' disease and a related but distinct syndrome, pontiac fever, are caused by infection with *Legionella pneumophila.*

Legionnaires' disease is a pneumonia which may be mild or severe. In severe cases the lung disease is progressive, complicated by gastrointestinal, nervous system, and kidney involvement, and may end in respiratory failure and death.

Pontiac fever is a short, self-limited illness, with influenza-like symptoms but no pneumonia.

Source of Infection. *Legionella* is ubiquitous in raw water systems, and infection is acquired by inhaling contaminated aerosolized water from air-conditioning cooling towers, condensers, shower heads, saunas, hot tubs, and drinking water systems. The number of cases attributable directly to drinking water is unknown. An estimated 250,000 cases of pneumonia are caused by *Legionella* each year in the United States.

The disease is not passed from person to person, and 80 percent of cases are isolated and sporadic. However, outbreaks have occurred in hotels, hospitals, and office buildings with a common air conditioning unit. The most severe cases have involved the elderly, and persons with AIDS or other immunocompromising illness. Definitive sources of infection are often difficult to determine, but water is usually strongly implicated.

Treatment. Erythromycin is the most effective antibiotic against *Legionella*.

Prevention. Air-conditioning units, shower heads, humidifiers, and other aerosols should be cleaned thoroughly, and water in distribution systems should be hyperchlorinated.

CHOLERA

Cholera is caused by infection with the organism, *Vibrio cholera*. Nontoxigenic strains may cause sporadic cases of diarrhea, whereas enterotoxin-producing serotypes cause epidemics. One toxigenic strain 0139 was responsible for recent widespread epidemics in India and Bangladesh. In the United States, cases occur in travelers who return from developing countries, and shellfish from the Gulf Coast of Louisiana and Texas are the source of sporadic cases.

Symptoms. A painless, severe watery diarrhea and vomiting are the characteristic manifestations of cholera in its severe or *gravis* form. Abdominal pain and fever are absent. The stools resemble "rice water" and are colorless. Patients rapidly become dehydrated and pass into shock, if fluids are not replaced. Children may have convulsions and lose consciousness.

Less than one in 20 people infected with *V. cholera* develop cholera gravis. The majority have no symptoms or suffer only mild to moderate diarrhea.

Diagnosis. Stool cultures and serological antibody tests will confirm the diagnosis in a suspected case.

Treatment. Fluids to correct dehydration and electrolyte imbalance and antibiotics are the mainstays of therapy.

Prevention. 1) Chlorination or boiling of water kills the organism and prevents its transmission.

2) Shellfish from the Gulf Coast should be cooked before eating.

3) Personal hygiene practices including hand washing should be observed.

4) Vaccines are of limited value. No country requires cholera vaccine for entry, and the World Health Organization does not recommend vaccination for travel to cholera-infected countries.

5) Cholera is a notifiable disease, internationally.

The chlorination of our public drinking water supplies has virtually eliminated cholera from the United States, and only occasional cases are reported, mainly from Gulf States. Other waterborne bacterial infections have also been drastically reduced. The majority of diarrheal illness attributed to drinking water is now caused by parasites, including *Cryptosporidium* and *Giardia,* and viruses.

Water other than drinking water, such as swimming pools, rivers, beaches, and fish tanks, may be contaminated with organisms that can cause disease by skin or mucous membrane contact. *Pseudomonas* causes "swimmer's ear infection." *Nontuberculous mycobacteria* can cause a skin infection, "swimming pool granuloma," or a throat and lung infection with lymph gland inflammation. *Leptospira* is a spirochete excreted in the urine of wild and domestic animals. Swimmers or waders in pools contaminated by animal urine may contract *Leptospirosis,* an influenza-like illness complicated by jaundice, kidney failure, meningitis, and eye swelling.

PREVENTING SPREAD OF WATERBORNE INFECTIONS

Young children in day care centers and patients in institutions and hospitals are particularly vulnerable to the spread of enteric bacteria, viruses, and parasites that cause diarrheal illness. When children and patients who are not toilet trained or are incontinent of feces are cared for in these centers, the risk of fecal contamination of the environment is particularly high.

Pathogenic organisms can be transmitted by fecal contamination or orally, either directly (by person-to-person spread) or indirectly (by toys, toilet seats, furniture, and food). When staff in centers or institutions care for incontinent children and patients and also prepare or serve food, the risk of enteric disease transmission is magnified. In some diseases, e.g. hepatitis A, the symptomatic illness occurs primarily among adult contacts of infected, asymptomatic children. Parents, nurses and staff personnel need to be aware of the epidemiology or spread of these diseases and the necessity for care in handling contacts and personal hygiene.

The single most important preventive measure to minimize fecal or oral transmission of diarrheal disease is **frequent hand washing.** Cleaning and disinfection of toilet seats, toys, furniture, and kitchen counters are other important precautionary measures in child care centers, institutions, and in the home.

American Academy of Pediatrics. 1994 Red Book. Report of the Committee on Infectious Diseases. Elk Grove Village, Illinois. AAP, 1994.

Baron EJ, Peterson LR, Finegold SM. Bailey & Scott's Diagnostic Microbiology, 9th ed. St Louis, Mosby, 1994.

Feigin RD, Cherry JD. (Eds). Textbook of Pediatric Infectious Diseases, 2nd ed. Philadelphia, WB Saunders, 1987.

Hoeprich PD, Jordan MC, Ronald AR. (Eds). Infectious Diseases, 5th ed. Philadelphia, JB Lippincott, 1994.

Keating JP, Dodge PR; and Furth S, Oski FA. Water intoxication as a potential etiology of seizures in young infants. AJDC Sept 1993;147:932-3.

Kuroki S et al. *Campylobacter jejuni* strains from patients with Guillain-Barre syndrome. Ann Neurol 1993;33:243-247.

Mandell GL, Douglas RG Jr, Bennett JE. (Eds). Principles and Practice of Infectious Diseases. 3rd edition. New York, Churchill Livingstone, 1990.

Millichap JG. Environmental Poisons in Our Food. Chicago, PNB Publishers, 1993.

Millichap JG. Ed. Progress in Pediatric Neurology II. Chicago, PNB Publishers, 1994.

Shulman ST, Phair JP, Sommers HM. (Eds). The Biological and Clinical Basis of Infectious Diseases, 4th ed. Philadelphia, WB Saunders, 1992.

U.S. Centers for Disease Control. Waterborne disease outbreaks, 1986-88, 1989-90, and 1991-92. Morbidity and Mortality Weekly Reports, 1990-93;39-42.

U.S. Environmental Protection Agency. Comparative health effects assessment of drinking water treatment technologies. Washington, DC, Office of Drinking Water, 1989.

CHAPTER **3**

WATERBORNE VIRAL DISEASES

Gastroenteritis and infectious hepatitis are the most common diseases caused by virus contamination of drinking water. Viruses involved in the waterborne transmission of infectious diseases are those that multiply in the intestine and are excreted in the feces of infected patients. These so-called "enteric viruses" enter the raw water supplies by way of sewage discharges. They are generally destroyed by adequate water treatment processes.

In developing countries and under certain conditions in the United States, water sources may be heavily polluted and the water treatment may be insufficient to eradicate all virus particles. The ingestion of contaminated water is known to have produced explosive outbreaks of viral gastroenteritis and hepatitis. It is also suspected of contributing to sporadic cases and endemic enteric viral disease.

Virus Detection. The examination of water samples for viruses takes a minimum of two weeks and the tests are not completely reliable. Bacteriological monitoring is used as the conventional indicator of potable water safety, but viruses are more resistant to water treatment processes than bacteria and may escape detection. Enteric viruses may be present in drinking water without signs of bacterial pollution. Viral bacteriophages that infect enteric bacteria are under investigation as indicators of enterovirus contamination.

The recycling of wastewater for domestic use, now employed in some States with water shortages, may increase the risk of virus contamination. One large beer producer in Southern California, with serious doubts about the purity of the reclaimed water used in manufacture of its product, has filed a lawsuit in Los Angeles to stop the practice of recycling sewer water. More sophisticated tests for the rapid determination of viral and parasitic contamination of potable water supplies might improve standards of safety and lessen consumers' concerns. Since 80 percent of beer is water, breweries may have to conduct their own tests for viral, bacterial, and parasitic, as well as chemical contamination, so that jokes about "beer made from sewer water" can be dispelled.

VIRAL GASTROENTERITIS

Norwalk virus, Norwalk-like agents, and hepatitis E virus account for approximately 5 percent of outbreaks of waterborne disease attributed to contaminated drinking water. All of these enteric organisms are included in the term, "caliciviruses," and some are named for the site of the outbreak where they were first identified.

Since the virus can be detected only by special laboratory testing, such as electron microscopy and antibody assays, the diagnosis is often missed or presumed from the clinical course. Outbreaks are frequently classified and reported as "acute gastroenteritis of unknown etiology."

Symptoms. The disease symptoms are diarrhea, vomiting, and stomach cramps, often complicated by influenza-like complaints, including fever, headache, malaise, and muscle aches. The infection lasts only 1 to 4 days and is self-limited. Excretion of the virus in the stools may continue for a few days after recovery. All age groups are affected, but children less than 4 years of age are most susceptible.

Source of Infection. Outbreaks of viral gastroenteritis occur after ingestion of contaminated water or foods like shellfish and salads. Transmission in close communities, such as child care centers, is person to person and by oral contact with infected feces. Symptoms begin within 1 to 4 days.

Diagnosis. For sporadic, self-limited cases, the virus is rarely identified. In community outbreaks, special research laboratories are involved in the diagnosis of the virus strain.

Control. The spread of infection can be prevented by careful hand washing, exclusion of infected persons or convalescent carriers from food handling, and isolation of sick young children.

Rotavirus and enteric strains of *adenovirus* are additional causes of viral gastroenteritis. Unlike Norwalk virus, they are rarely attributed to the ingestion of contaminated drinking water. Most cases result from contact with infected humans, and spread by the fecal-oral route within institutions is common.

Rotavirus is the most common cause of infectious diarrhea acquired in children's hospitals and in children attending day care centers. Control precautions to limit person to person contact are important because excretion of the virus in stools persists after recovery. Surfaces of toilet seats and areas for food preparation should be washed or disinfected frequently.

Other enteric viruses and diseases that may be transmitted via drinking water contaminated by human feces are included in Table 3-1.

HEPATITIS A INFECTION

Symptoms. Hepatitis A infection presents with fever, jaundice, loss of appetite, nausea, and a feeling of malaise. The disease is acute in onset and self-limited. It does not become chronic.

Source and Method of Spread. Outbreaks occur from the ingestion of drinking water or shell fish contaminated by human sewage. Transmission is by stool contamination of fingers - the fecal-oral route.

Table 3-1. Viral Agents of Waterborne Disease.

Virus	Disease
Norwalk-like agents	Gastroenteritis
Hepatitis A	Infectious hepatitis
Rotavirus	Gastroenteritis
Enteric Adenovirus	Gastroenteritis
Polioviruses	Poliomyelitis
Coxsackie viruses	Aseptic meningitis
Echoviruses	Aseptic meningitis

Endemic in developing countries, hepatitis A spreads in households and child care centers in the United States, sometimes in epidemic proportions. It also occurs among young adults who are intravenous drug users.

Adult contacts of infected children become jaundiced (icteric) and sick, while young infants harboring the virus are often asymptomatic and appear well. Infants in diapers are an important source of contagion. Virus-infected stools are passed for 1 to 2 weeks before the onset of symptoms, and the risk of transmission of infection lessens during the week after the jaundice develops. The incubation period - time from first contact to the onset of jaundice - is an average of 3 to 4 weeks.

Diagnosis. Diagnosis is made by the clinical manifestations and course of the illness. It may be confirmed by tests for the development of specific antibodies in serum.

Precautions and Control. Ill patients should observe particular care in personal hygiene and avoid food preparation for others for one week after the onset of jaundice. Children or teachers with acute symptoms should not return to school for one week. Infants with symptoms should be excluded from day care centers. Hepatitis A virus may survive on changing tables and on other objects in the environment for weeks, and disinfection of surfaces and toys is important in the control of spread of the infection.

Immune globulin prophylactic injections are prescribed by the physician for household and sexual contacts, and for employees and children in child care centers with an identified case, and their household contacts. It is also recommended for personnel in close contact with patients in a hospital or institution when an outbreak occurs. Travelers to developing countries who may be susceptible should receive immune globulin. They should avoid potentially contaminated water that is not carbonated and bottled and should not eat cold uncooked foods. Vaccines have been developed but are not yet licensed for use in the United States.

Hepatitis B. Whereas hepatitis A virus infections are generally self limited and free from serious complications, hepatitis B virus causes a chronic infection with serious liver disease and the risk of liver carcinoma in later life. Unlike hepatitis A, hepatitis B is transmitted through blood or body fluids such as semen and saliva, not via contaminated drinking water nor by the fecal-oral route.

AAP 1994 Red Book: Report of the Committee on Infectious Diseases. 23rd edition. Elk Grove Village, Illinois,

American Academy of Pediatrics, 1994.

EPA. Comparative Health Effects Assessment of Drinking Water Treatment Technologies. Washington DC, Environmental Protection Agency, Office of Drinking Water. 1989.

Mandell GL, Douglas RG Jr, Bennett JE. (Eds). Principles and Practice of Infectious Diseases, 3rd edition. New York, Churchill Livingstone, 1990.

World Health Organization. Guidelines for Drinking-Water Quality. Vol 2. Health Criteria and Other Supporting Information. Geneva, WHO, 1984.

CHAPTER 4

WATERBORNE PARASITIC DISEASES

P arasitic infections are the most frequently identified causes of disease attributed to contaminated drinking water in the United States. *Giardia lamblia* infection was responsible for 20 (21%) outbreaks and 1600 (3%) cases, and *Cryptosporidium* accounted for 3 (3%) outbreaks but 16,000 (35%) cases reported to the Centers for Disease Control, Atlanta, in the period 1986-92. In the year following this report period, an outbreak of acute gastrointestinal disease caused by *Cryptosporidium* in the City of Milwaukee affected close to half a million residents and caused 100 deaths.

Unlike disease-related bacteria that are controlled by chlorination, parasitic cysts are resistant and require filtration of raw water for their removal.

The inadequacy and breakdown of a municipal water filtration plant were the reasons for the recent outbreak of *Cryptosporidiosis* in Milwaukee.

A list of parasitic protozoa and helminths (worms) and diseases that may be transmitted in contaminated drinking water and via human feces is shown in Table 4-1.

TABLE 4-1. Waterborne Parasites and Diseases.

Parasite	*Disease*
Protozoans:	
Giardia lamblia	Giardiasis
Cryptosporidium	Cryptosporidiosis
Blastocystis hominis	Diarrhea
Balantidium coli	Dysentery
Entamoeba histolytica	Amoebic dysentery
Helminths:	
Ascaris lumbricoides	Round worms
Trichuris trichiura	Whipworms

PROTOZOAN INFECTIONS

GIARDIASIS

Giardiasis is the disease caused by *Giardia lamblia*, a protozoan parasite with cysts that infect the small intestine of humans and domestic and other animals.

Symptoms. Acute and chronic forms of the disease are recognized. Acute giardiasis presents as a sudden onset of explosive, watery diarrhea with abdominal pain. Chronic giardiasis is an extended illness characterized by foul-smelling diarrheal or soft, fatty stools associated with bloating of the stomach, flatulence, poor appetite, and weight loss.

Source and Spread of Infection. Community-wide epidemics result from contaminated water supplies. Person-to-person transmission by hand-to-mouth transfer of cysts in feces of infected contacts is common in institutions for the mentally retarded, and in child care centers. Patients with immune deficiencies and especially those with cystic fibrosis are particularly susceptible.

Cyst excretion may persist for months, and cysts remain viable and infectious in a moist environment for at least three months. Many persons with cysts found in stools are without symptoms but still capable of passing on infection.

Diagnostic Tests. Laboratory microscopic examination of a single stool specimen will detect up to 75% of *Giardia* infections. Three specimens collected consecutively will increase test sensitivity to 95%. Stool collection kits should be obtained for preservation of specimens provided from home.

Treatment. The drug Quinacrine is usually prescribed by physicians for the treatment of acute or chronic infections. Metronidazole is an alternative medication in adults, and furazolidine available in liquid suspension is sometimes preferred for children. Those found to be cyst carriers but without diarrhea are generally not treated.

Control. Prevention of waterborne outbreaks depends on adequate filtration of municipal surface water supplies. Chlorine in the concentrations used to kill bacteria in drinking water is not effective against *Giardia* and other parasitic cysts.

Boiling of water eliminates cysts in potentially contaminated water obtained from streams and camp grounds or in developing countries.

In child care centers, hand washing and careful personal hygiene should be emphasized, and infected persons should be excluded until diarrhea is controlled.

CRYPTOSPORIDIOSIS

Cryptosporidiosis is a diarrheal illness caused by the protozoan parasite, *Cryptosporidium.* It is emerging as a major cause of epidemic outbreaks of waterborne gastrointestinal infections in the United States and worldwide.

Symptoms. The most common presenting symptoms are frequent, frothy or watery bowel movements, low-grade fever, abdominal cramps, poor appetite, and loss of weight. In general the infection is self limited, with recovery expected within one to two weeks. In immunocompromised patients with AIDS, the diarrhea becomes chronic and is complicated by dehydration, poor nutrition, and sometimes even death.

Source. *Cryptosporidium* is transmitted from farm animals or pets to humans. Waterborne outbreaks result from contamination of surface water by animal wastes and failure of municipal filtration systems. Person-to-person transmission occurs in child care centers and institutions. The parasite oocyst is resistant to chlorine, and efficient filtration technologies are essential.

Diagnosis. Physician awareness of the symptoms and early detection of the parasite in initial stages of an outbreak should lead to isolation of the source, and a prompt warning to consumers to boil or avoid the water from the faucet. Delay in diagnosis in persons who sought medical help was one reason for the widespread extension of the recent Milwaukee epidemic.

Parasitic infections are often misdiagnosed as "viral" gastroenteritis or "intestinal flu," and the physician may fail to request the special laboratory procedures required to detect the ova and parasites in a stool specimen. Oocysts are small, about 5 micrometers in diameter, and the microscopic examination of the stool may require concentration of the cysts and special staining techniques. At least two and preferably three stool specimens should be examined before the diagnosis can be excluded.

Treatment. No specific treatment is available. Enteric precautions are recommended for hospitalized patients to prevent spread among contacts.

BLASTOCYSTIS DIARRHEA

Symptoms. The diarrhea associated with the protozoan parasite *Blastocystis hominis* is explosive and watery, and accompanied by bloating and flatulence. It often occurs immediately after eating. There is no abdominal pain or nausea, and the stools are not bloody.

The relation of *Blastocystis* infection to gastrointestinal disease is controversial, some having considered the parasite a coincidental finding.

However, an increased interest in this infection is now emerging because of the frequency of stools positive for *Blastocystis* among immunocompromised patients with chronic, incapacitating diarrhea.

Source of Infection. Animals, including pigs and poultry, are infected, and transmission to humans is probably indirect, via contamination of food and water by animal feces and wastes, directly by person-to-person contact and the fecal-oral route, or via infected food handlers in restaurants.

Blastocystis hominis has been recovered from 25% or more of stool samples examined for ova and parasites in some surveys. The asymptomatic carrier is well documented. It may occur in association with other parasites, such as *Entamoeba histolytica,*. However, the pathogenic significance of *Blastocystis* in some cases of chronic diarrhea is accepted, and a satisfactory response to treatment with metronidazole is reported.

Diagnosis requires stool examinations using a freshly passed specimen or a preservative and special stains. The laboratory should be asked to look specifically for *Blastocystis.* Failure to recognize and treat this parasite as a pathogen, particularly in patients with immunocompromising illnesses, can lead to unnecessary, costly, and unpleasant diagnostic procedures such as colonoscopy and barium radiographs.

Treatment. A 5 or 10 day course of metronidazole is usually effective. Iodoquinol is an alternative agent sometimes prescribed.

BALANTIDIASIS

Balantidiasis is a diarrheal illness caused by the ciliated protozoan parasite, *Balantidium coli. B. coli* is the largest protozoan known to infect humans. The ciliated trophozoite stage measures 50-200 mcm in length. Its surrounding membrane is covered by hundreds of tiny cilia that facilitate movement and penetration of the intestinal wall. The cyst stage, also excreted in the feces, is very resistant and can remain in the environment for months.

Symptoms. The acute infection is characterized by nausea, vomiting, abdominal discomfort, and diarrhea with blood and mucus in the stools. In some cases, the diarrhea is intermittent and in others, it becomes chronic, with ulceration of the large intestine, and rarely, bowel perforation and peritonitis. Many carriers of the parasite are without symptoms.

Source and Spread of Infection. Pigs have a particularly high rate of infection and are the main reservoir of *B. coli* for humans. Infection is transmitted indirectly through fecally contaminated water or food, or directly from hand to mouth.

Humans are generally resistant to *B. coli* infection, but persons with poor nutrition or immunosuppression are particularly susceptible. Outbreaks may occur in mental institutions where personal hygiene among residents is often lacking.

Diagnosis. The rapidly motile trophozoites can be detected microscopically in infected stools, provided the specimen is fresh and is examined promptly. Cysts are detected less often. Three specimens may be necessary for a positive diagnosis. Examination of scrapings taken from bowel ulcers may be required in some chronic cases.

Treatment. The antibiotic tetracycline is prescribed in infected adults. Iodoquinol and metronidazole are alternative treatments. Tetracycline causes permanent staining of teeth in young children and is contraindicated under nine years of age.

Control. Avoid water and food that may be contaminated by pig wastes, and observe care in sanitary disposal of human feces and in personal hygiene.

AMEBIASIS

Amebiasis is an intestinal disease caused by the protozoan parasite, *Entamoeba histolytica.*

Symptoms. Acute amebic dysentery is a colitis presenting with diarrhea containing blood and mucus, abdominal cramps, and pain on defecation. Infection occasionally spreads to the liver, causing an amebic abscess. Rarely, the lung, brain, genitalia, and kidneys are involved.

More commonly, amebiasis presents in a less severe form, with mild intestinal symptoms, including abdominal distension, flatulence, and intermittent diarrhea with abdominal cramps.

Source and Spread of Infection. Amebiasis is spread by cysts that are transmitted from person-to-person via contamination of water and vegetables or by direct fecal-oral contact. Asymptomatic cyst carriers are a major source of infection. Cysts are resistant to disinfectants and they can survive for weeks or months in a moist environment.

The prevalence of infection is as high as 50 percent in underdeveloped countries and 5 percent in the United States. The incidence of infection is higher in immigrants from endemic areas, travelers returning from underdeveloped countries, in institutions for mentally retarded, and in homosexuals. Risk factors for increased severity of infection include young age groups, particularly newborns, pregnancy, use of corticosteroids, malignancy, undernutrition, and AIDS.

Overcrowding and poor sanitation and hygiene have resulted in major waterborne outbreaks, including the Chicago Exposition in 1933, and the Singer Sewing Machine Plant in Indiana in 1950.

Diagnosis. Microscopic examination of serial stool specimens should show amebic cysts or trophozoites. Immunologic tests on serum are helpful in the diagnosis of liver abscess.

Treatment. Trophozoites and cysts have to be eliminated from the intestine and bowel wall. Metronidazole followed by iodoquinol are prescribed for amebic dysentery and abscess, and for cases of mild to moderate diarrhea. Iodoquinol is effective in the elimination of cysts in carriers without symptoms of infection.

Emetine injections, an older form of treatment, is now used only in cases of liver abscess or invasive amebiasis that have failed to respond to metronidazole.

Prevention. Amebic infection can be prevented by avoiding fecal contamination of water and food. Water is a prime source of spread of infection, and amebic cysts are not killed by chlorine or iodine disinfection. Only boiling of water will eradicate amebae.

Foods most commonly contaminated are fresh vegetables, especially lettuce. In endemic, tropical areas and at times of outbreaks of amebiasis, vegetables should be avoided or thoroughly cooked before eating. Lettuce should be washed with soap detergent and soaked in vinegar. Salads in hotels and restaurants, even in developed countries, can be suspect if prepared by food handlers who are recent immigrants from endemic locations.

Hand washing after defecation, appropriate disposal of human waste, boiling of drinking water or use of carbonated, bottled water, and adequate cleaning or cooking of vegetables will prevent the spread of amebiasis under most circumstances. The avoidance of sexual practices that facilitate fecal-oral contact, or use of condoms can reduce the risk of sexual transmission.

PARASITIC WORM INFECTIONS

ASCARIASIS (ROUNDWORM)

Ascariasis is the clinical syndrome caused by infection with the roundworm, *ascaris lumbricoides.*

Symptoms. Some patients report ill-defined gastrointestinal symptoms such as nausea and abdominal discomfort. Usually, there are no symptoms and the patient reports having vomited or passed a large worm in the stools. Complications include the occasional occurrence of an acute intestinal obstruction, appendicitis, jaundice, and pneumonia.

Source and Life Cycle. Infection results from the ingestion of ascaris eggs in water or in uncooked vegetables contaminated by soil. Ascariasis is most common in the tropics and in travelers returning from underdeveloped countries. It is endemic in areas with poor sanitation and water supplies contaminated by human feces.

The life cycle of *Ascaris* is 4 to 8 weeks. After ingestion of the ascaris eggs in water or food, it takes about 2 months for the adult worm to develop and start laying eggs in the intestine. If untreated, adult worms can exist in the small intestine of humans for months. The adult female worms produce thousands of eggs daily which are excreted in the stools. To become infectious the eggs must first incubate in soil for 2 weeks.

The eggs ingested by humans will hatch larvae which penetrate the wall of the intestine and pass by the blood stream to the liver and lungs. From the lungs they ascend the bronchial tree to the trachea and pharynx, where they are swallowed, and become mature worms in the small intestine. Symptoms may result from irritation or obstruction by the larvae in the lungs, liver, or the intestine.

Diagnosis. If a worm is passed by the patient, it should be collected if possible for inspection by the physician. A stool specimen is sent to the laboratory for detection of ova by microscopic examination.

Treatment. Treatment with either pyrantel or mebendazole is recommended by the physician for cases with or wihout symptoms. If left untreated there is a danger of complications, sometimes requiring surgery.

Prevention. Vegetables eaten in under-developed countries, where human feces may be used as fertilizer, must be thoroughly cooked or avoided. Sanitary disposal of feces and disinfection of children's play areas and toys are important precautions and control measures.

TRICHURIASIS (WHIPWORM)

Trichuriasis is the diarrheal illness caused by infection with whipworm, *Trichuris trichiura,*

Symptoms. Trichuriasis presents with abdominal pain, diarrhea, blood in the stools, and pain on defecation. The worms, 3 to 5 cm in length, penetrate the lining of the large bowel.

Source and Spread. The disease, prevalent in the tropics, is associated with poor sanitation. It has been reported in the southeastern areas of the United States, usually among immigrants from underdeveloped countries. The eggs become infectious after 10 days incubation in soil and are ingested in water or food. Whipworm is not transmitted from person to person.

Diagnosis. Eggs may be detected in the stool on microscopic examination.

Treatment. Mebendazole is usually prescribed.

Prevention. Proper sanitation and disposal of human feces.

The roundworm, *Strongyloides,* and hookworm, *Necator americanus,* are usually transmitted by infectious larvae that penetrate the skin while walking bare foot. They occur predominantly in rural, tropical areas where sanitation is poor and soil contaminated by human feces is common. These infections are generally not transmitted by contaminated drinking water.

AAP. 1994 Red Book: Report of the Committee on Infectious Diseases. 23rd Edition. Elk Grove Village, Illinois, American Academy of Pediatrics, 1994.

Environmental Protection Agency. Comparative Health Effects Assessment of Drinking Water Treatment Technologies. Washington, D.C., Office of Drinking Water, 1989.

Feigin RD, Cherry JD. Textbook of Pediatric Infectious Diseases, 2nd edition. Philadelphia, WB Saunders, 1987.

LeMaistre CA, Sappenfield R, Culbertson C et al. Studies of a water-borne outbreak of amoebiasis: South Bend, Indiana. Am J Hyg 1956;64:30-45.

Mandell GL, Douglas RG Jr, Bennett JE. (Eds). Principles and Practice of Infectious Diseases. 3rd edition. New York, Churchill Livingstone, 1990.

Select Committee. Amoebiasis outbreak in Chicago: Report of a special committee. JAMA 1934;102:369.

CHAPTER 5

LEAD IN OUR DRINKING WATER

L ead in drinking water contributes between 10 and 20 percent of the total environmental exposure to lead in young children. Water as a source of lead poisoning may increase in relative importance as abatement programs reduce exposure from lead-based paint and urban soil and dust.

The U.S. Environmental Protection Agency (EPA) estimated in 1986 that some 40 million Americans were drinking water that contained potentially hazardous levels of lead. Until the year 1993, the EPA limit for lead levels in drinking water was 50 ppb (parts per billion). The new limit has now been set at 15 ppb.

Recent EPA tests of water from household taps failed to meet the new Federal standards in almost 20 percent of 660 larger communities, nationwide. Ten public water suppliers in the State of Illinois, including 7 in Chicago suburbs, were included in the list of geographic areas with consumers at risk. Adding phosphate to the main water supply to control corrosion of plumbing is one solution to the problem. The replacement of lead service pipes would be a more costly alternative.

RISK FACTORS FOR LEAD IN WATER

Lead levels in the drinking water are likely to be highest under the following conditions:
- If the home or water system has lead pipes.
- If the home has copper pipes with lead solder, brass faucets, or brass fixtures, and
- If the home is less than five years old, or
- If the drinking water is soft, or
- If water sits in the pipes for many hours.

A 1986 law banned the use of leaded solder on pipes that carry drinking water, but the EPA enforcement of this ban is limited. The problem is found in houses that are either very old or very new.

Plumbing installed before 1930 is most likely to contain lead. Lead was also used for the service connections, goose necks, and water meters that join residences to public water supplies. This practice ended only recently in some localities. Copper pipes have replaced lead pipes in most residential plumbing. Much of household plumbing consists of copper pipes connected by lead-containing solder. The solder is the major cause of lead contamination of household water in the United States today.

Lead enters the water after the water has left the local treatment plant or private well. Corrosion, a reaction between the water and the lead pipes or solder, is the most common cause. All kinds of water may be contaminated but "soft" water is more likely to facilitate corrosion than "hard" water.

Contrary to popular belief, the newer the home, the greater the risk of lead contaminated water. The newer the plumbing installation, the greater the degree of lead leaching from the pipes, especially when the water is soft, acidic, and corrosive. Older pipes have developed a protective coating of mineral deposits on the inside. This coating insulates the water from the solder and decreases the danger of leaching lead. *Buildings less than five years old may have the highest levels of lead in the drinking water!*

LEAD LEVEL WATER STANDARDS

Until recently, Federal standards limited the amount of lead in water to 50 parts per billion (ppb), measured at free-flowing taps. In the light of new health and exposure data, the EPA has recently tightened this standard to a 15 ppb first flush "action level" at the tap. The EPA also requires all large water supply systems to optimize corrosion control for lead.

If the level of lead in the household water is in the area of 15 ppb or higher, it is advisable to take measures that will reduce the lead level to safer amounts, especially when there are young children in the home. Tap water should not be used to prepare infant formula.

The EPA estimates that more than 40 million U.S. residents use water that contains lead in excess of 20 ppb, when care is not taken to discard the first flush water obtained from the faucet. Steady exposure to 20 ppb in drinking water would contribute between 2.5 and 3.5 mcg/dL to a child's blood lead level.

It is estimated that the new EPA drinking water regulations will result in a reduction of the average blood lead level from 5.3 to 4.7 mcg/dL, in those children not exposed to paint or soil contamination hazards. Approximately half a million children with hazardous levels will have their blood lead levels reduced below 10 mcg/dL, and lead exposure will be minimized for millions of Americans.

SOURCES OF LEAD

The sources of lead exposure and poisoning vary with age and are different in children and adults and even in infants, toddlers, and older children. The one source that is common to all ages is drinking water. Water is the chief cause of lead poisoning in young infants.

Infants have been exposed to lead in the baby formula when inadvisably prepared or reconstituted using water from the hot faucet. Formula preparation with lead contaminated water accounted for 9 of 50 infantile cases of lead poisoning reported in a pediatric journal in 1992. These data support recent recommendations to initiate lead screening in infants at 6 months of age. Hot water storage cisterns may be lead lined, and lead will leach from solder in pipes more freely into hot than cold water, especially if the water is soft and at an acid pH.

Toddlers are prone to "pica," a craving for unnatural items of food. They also indulge in hand-to-mouth activity. The ingestion of paint chips and house dust is the most common source of lead poisoning in children aged 18 to 30 months, accounting for 86% of cases in this age range in one recent report.

Lead containing paint is present in three quarters of all private dwellings built in the U.S. before 1978, and housing and apartments built before 1950 are at even greater risk of harboring interior lead-based paint. Multilayered chips of old lead pigment paints may contain 20,000-100,000 mcg Pb/cm^2. House dust in old houses may average from 600-3,000 mcg Pb/g, and garden soil can contain 2,000-16,000 mcg Pb/g. These are the major sources of exposure to lead in young children in the United States.

The estimated number of children under 6 years of age exposed to various lead sources are compared in Table 5-1.

TABLE 5-1. Numbers of Children under Six Years Exposed and having Elevated Blood Lead Levels.

Lead Source	Children Exposed	Blood Lead > 10 mcg/dL	
		No.	Percent
Paint	12 million	2 million	17
Soil and Dust	12 million	unknown	-
Drinking Water	30 million	1 million	3

Adapted from the EPA Strategy for Reducing Lead Exposures, February 21, 1991.

Older Children and Adolescents may be exposed to lead-containing fumes from burning painted wood or casings of storage batteries used for heating living quarters in socioeconomically deprived areas of a city. They may also risk poisoning by the hedonistic sniffing of leaded gasoline. Aerosol exposure is especially hazardous, since approximately 40 percent of inhaled lead is absorbed.

The banning of lead in gasoline has resulted in a substantial decline in the blood lead levels of the entire U.S. population between 1976 and the present. The enforced removal of lead from soldered cans containing food and beverages has also limited the danger of another potential source of lead poisoning in children and adults.

Adults are exposed to occupational sources of lead. Welders, miners, pottery makers, ship and bridge builders, auto workers, printers, house painters and renovators, demolition workers, and employees in gun clubs and firing ranges are particularly vulnerable. Estimates place one million American workers at risk of lead exposure from their occupational environments.

Before the Industrial Revolution the total body content of lead was about 2 mg. In industrialized societies today, this figure has increased 100 times. Of 200 mcg of lead ingested each day, about 5 to 10 percent is absorbed, and the amounts are even higher in children. In malnourished children with deficiences of iron, calcium, magnesium and zinc, the absorption and potential for toxicity from lead exposure are increased. Diets low in iron and calcium are prevalent among children in low income groups. Of all dietary factors involved, iron deficiency is the most important nutritional problem that predisposes young children to lead poisoning.

INCIDENCE OF LEAD POISONING

Seventeen percent of children in the United States aged 6 months to 6 years are reported to have blood lead levels greater than 10 to 15 mcg/dL. The Agency for Toxic Substances and Disease Registry (ATSDR) has defined the threshold for neurobehavioral toxicity for lead at 10 mcg/dL. As sustained lead levels rise above this threshold, children exposed are at a progressively increased risk for future cognitive and behavioral abnormalities which can lead to school problems and learning deficits.

Chronic lead absorption is most prevalent in preschool children living in old, deteriorated housing. Lead-containing paint and the highest levels of lead in surface soil and interior household dust are found in and around this type of housing. Plumbing in older housing is likely to contain lead in pipes and solder, and drinking water would continue to pose a problem of lead exposure after the abatement of the lead contaminated paint and dust.

SYMPTOMS AND SIGNS

The symptoms and signs of lead poisoning are diverse and often subtle. The term *silent epidemic* has been used to describe the discovery of children suffering from learning and behavior disorders caused by exposure to lead during infancy and even before birth. The recognition of early clinical manifestations of lead poisoning and a knowledge of environmental risk factors are important to all consumers, and especially to parents of young children who are most sensitive to the toxic effects of lead.

All systems of the body may be affected, but the effects of lead on the nervous system are of most concern.

Nervous System Effects. Central nervous system symptoms and signs are vague at first and include irritability, incoordination, memory lapses, sleep disturbances, listlessness, headache, and dizziness. Symptoms of brain damage may occur with blood lead levels of 25 to 60 mcg BPb/dL or less in children.

The long-term effects of low dose lead exposure (10 to 25 mcg/dL blood lead levels) during early childhood have been correlated with impairment of academic success at a later age. Studies have shown a failure to graduate from high school, lower class standing, absenteeism, and dyslexia in adolescents exposed to lead at an early age. Deficits in vocabulary, fine motor skills, coordination, and reaction time have been reported.

The lead content of dentin in teeth shed at ages 6 and 7 years showed a positive correlation with neurobehavioral impairments as measured in young adulthood. Young people with dentin lead levels greater than 20 ppm (parts per million) had a markedly higher risk of dropping out of high school and of having a reading disability as compared with those whose dentin lead levels were less than 10 ppm. Exposure to lead even in children without obvious symptoms may have an important and enduring effect on brain functioning and learning.

Acute lead encephalopathy. More striking neurological manifestations of lead poisoning include vomiting, severe headache, convulsions, blindness, and loss of consciousness, symptoms of an encephalopathy. Acute encephalopathy is more common in children than in adults and carries a poor prognosis. The mortality rate is high and in children who survive, brain damage is frequent, with persistent neurologic sequelae including seizures, paralyses, and mental retardation.

Of 5 children with lead poisoning that I reported in 1952 in London, England, 4 had acute lead encephalopathy and all recovered. The fathers of these children were unwittingly responsible for the poisoning of their own offspring by painting their cribs with lead-based paint. The children suffered from pica, and they had ingested lead by chewing the paint on the crib rails. Only one of 20 cases of lead poisoning reported in the British medical literature in the previous 50 years was caused by drinking water.

Lead encephalopathy has become an infrequent complication of lead poisoning since the banning of lead in paint, the introduction of more stringent laws regarding lead abatement, and increased screening efforts to detect elevated blood lead levels in children at risk. A fatal case of poisoning from lead-based paint was reported in Wisconsin in 1990: a 28-month-old boy was brought to the hospital because of decreased appetite and lethargy. His parents stated that he had eaten paint chips. His blood lead level was 144 mcg/dL; levels as low as 70 have been fatal. The child fell into a coma and died within 26 hours of admission. An autopsy revealed swelling of the brain, paint chips in the intestine, and deposits of lead in the bones. The paint in the home was found to be chipping badly, and analysis showed it to be one third lead.

The previous report of a fatal case of poisoning from lead based paint in the United States was in the mid-1970s, an interval of 20 years between cases. In developing countries, however, the incidence of acute lead encephalopathy is higher; 10 cases were diagnosed in Saudi Arabian children, ages 8-48 months, during the years 1984-1988.

Spinal cord involvement. The spinal cord is affected in lead poisoning only rarely, producing a syndrome that resembles Lou Gehrig's disease (amyotrophic lateral sclerosis). Some studies have shown that patients with Lou Gehrig's disease report exposure to lead and mercury, the consumption of large amounts of milk, and participation in athletics and manual labor, factors that may increase susceptibility to lead-induced paralyses.

Peripheral nerve paralyses. The peripheral nervous system is involved more often in adults than children. Paralyses of the muscles supplied by the radial nerve result in weakness of finger extension and wrist drop. In adults leg paralyses are exceptional, but the converse is true in children who may present with weakness and wasting of the muscles of the feet or sometimes of the quadriceps.

Abdominal symptoms. Intermittent abdominal cramps or colic are associated with nausea, vomiting, loss of appetite, and constipation. The pain may simulate appendicitis and other surgical intestinal disorders that must be considered in diagnosis.

Other symptoms include anemia, kidney failure, and joint pains. A blue line on the gums is found in 20 percent of adults with chronic lead exposure but is infrequent in children.

Lead crosses the placenta and is responsible for spontaneous abortion and miscarraige. It also causes stunting of fetal growth and childhood stature by effects on growth hormone secretion. Lead exposure may result in chromosome anomalies and reduced sperm counts.

DIAGNOSIS OF LEAD POISONING

Symptoms are nonspecific and mimic other diseases. A high index of suspicion is important in making the diagnosis. Lead poisoning should be suspected in a child with pica, living in an old, deteriorated house, who presents with abdominal pain, vomiting, irritability, seizures, or anemia. The environment should be examined for sources of lead, and early diagnosis depends on prompt screening of blood for lead.

Blood lead level is the diagnostic screening test of choice. A level of 10 mcg Pb/dL or greater is unacceptable, and risks of toxicity and neurobehavioral effects increase in stages according to lead levels:
- Stage I, <10 mcg/dL, Accepted as safe.
- Stage IIA, 10 - 14 mcg/dL, Borderline zone
- Stage IIB, 15 - 19 mcg/dL
- Stage III, 20 - 44 mcg/dL
- Stage IV, 45 - 70 mcg/dL
- Stage V, >70 mcg/dL, Medical Emergency.

Medical follow-up and treatment strategies vary with the determination of the stage of poisoning. Patients show a varying susceptibility to lead and become symptomatic at different levels. A high lead screening test level is not pathognomonic of lead encephalopathy in a comatose child, and alternative diagnoses such as cerebral tumor and viral encephalitis must be considered before specific treatment is begun.

TREATMENT OF LEAD POISONING

The removal of the child from the source of lead and removal of the lead hazard from the environment are the cornerstones of therapy. Pregnant women should also evacuate the premises until the abatement of the lead paint has been completed. If drinking water is the source, bottled water should be substituted until the plumbing can be modified, or care should be taken to run the faucet until the water is cold before drinking. Details for the prevention of poisoning by lead in drinking water are outlined below.

Lead laws and lead abatement regulations in some States are very stringent, leading to costly and trying circumstances for property owners and realtors. Laws address disclosure of lead hazards in properties for sale, and contracts are required to warn buyers that older houses may contain lead. Buyers are recommended to obtain a lead inspection, including a water analysis for lead, before the purchase of an older home. The plumbing in some newer highrise apartments may also be suspect.

When a lead paint source of exposure is discovered, the premises should be professionally vacuumed, scrubbed with high-phosphate detergents, and then repainted with unleaded paint. Play in dirt adjacent to the housing should be avoided. The patient is tested periodically to determine the progress of lead levels, and other children in the household are also examined for evidence of plumbism (lead poisoning).

   Treatment Regimens.   Most children detected with increased blood lead levels are without symptoms and fit the classifications of Stages IIA or IIB. They are at risk for subtle and long-term, chronic effects on cognitive functioning. Parents should be informed of the effects of diet and nutrition on lead absorption, emphasizing the benefits of foods high in iron and calcium. Patients in Stages IV and V, and sometimes those in Stage III, require chelation therapy to facilitate the excretion of lead in the urine. Children in Stage III should receive a medical evaluation; those in Stage IV need both medical and environmental intervention; and Stage V requires immediate therapy.

   Chelating agents include CaNa Versenate (CaNaEDTA), BAL (dimercaprol), and Chemet (succimer). CaNaEDTA is employed in patients with blood lead levels above 45 mcg/dL and in those with acute encephalopathy. BAL is used to initiate chelation in patients with Stage V poisoning and blood lead levels above 70 mcg/dL.

   Succimer is preferred for patients with milder intoxications and blood lead levels above 45 and below 70 mcg/dL. It is not FDA approved for the prophylaxis of lead poisoning in areas of high risk exposure or for treatment of patients who are symptomatic or have blood lead levels of 70 mcg/dL or higher. It is administered by mouth and allows the option of outpatient management and lower health care costs. Succimer chelates are excreted in the urine, and adequate fluid intake is essential to maintain good urine flow. The drug has a "rotten egg" sulfur odor which is a disadvantage. Unlike CaNaEDTA, succimer does not induce acute zinc depletion, and it is more effective than CaNaEDTA in lowering the lead content of brain, kidney, and blood. Dosages and duration of treatments are individualized according to the patient's needs and the judgment of the physician.

Effective long term management of lead poisoning requires the cooperative efforts of health department personnel, social worker, psychologist, and pediatrician or family practitioner. Regular follow-up with blood lead determinations and attention to nutrition and hygiene are important. Pica and hand-to-mouth activity that predispose to plumbism may be benefited by behavior modification.

PROGNOSIS

Even Stage IIA intoxications, with lead levels of 10 to 14 mcg/dL, if sustained during early childhood, may result in subtle injury to the nervous system. Learning and behavioral problems can be manifested in later childhood and adolescence. After effects of poisoning are related to the degree and duration of lead ingestion and absorption. Survivors of acute encephalopathy may be mentally retarded and suffer from seizures and paralyses.

Although early diagnosis, and effective treatment can lower the risk of serious neurological and kidney damage, the prevention of exposure to lead in our environment is essential for the control of this "silent epidemic."

PREVENTION

Public Health Agencies in the United States have attempted to reduce the risks of lead exposure and poisoning by enforcing drinking water standards for lead, reducing the lead content of canned food, and in 1988, eliminating lead additives in gasoline. The lead in residential paints was banned by the United States Consumer Product Safety Commission in 1977, but lead-based paint and lead-contaminated dusts remain the primary sources of lead exposure for children.

Lead remains in approximately 74 percent of all private dwellings built before 1978, and housing built before 1950 is at even greater risk of having interior lead-based paint. Methods of abatement of lead in paint and dust in homes, prevention of exposure to lead in foods, and occupational risks are detailed in a previous publication, "Environmental Poisons in Our Food."

Prevention of lead exposure via household drinking water is the responsibility of both the government and the consumer. The government, through the EPA, and under the authority of the Safe Drinking Water Act and its amendments, has recently tightened the water lead level standard to a 15 ppb first flush "action level" at the tap (i.e. the water that immediately flows when a tap is first opened). The EPA in 1986 required all public water systems to optimize corrosion control of lead and to use lead-free pipes, solder, and flux in new installations or repairs. These provisions also apply to residential or non-residential facilities connected to a public water system (i.e. any system that supplies water to 25 or more people).

CONSUMER LEAD-FREE WATER TIPS

Consumer Steps to Reduce Lead Exposure from Drinking Water include the following:

• Refrain from drinking water that has been in contact with the plumbing for more than six hours. "Flush" the cold water faucet until the water is cold before drinking. (The water that flows after flushing will not have been in extended contact with lead pipes or solder, except in some high-rise buildings with large diameter supply pipes).

• Never cook with or drink water from the hot-water tap. Hot water dissolves more lead more quickly than cold water.

• Have your water tested for lead. Testing costs between $20 and $100. Testing is especially important in high-rise buildings where flushing may not result in reduced levels of lead. Home test kits may soon be available from *HybriBet Systems, Inc., P.O. Box 1210, Framingham, MA 01701, Tel. 1-800-262-LEAD.*

• Ask your public water supplier if the water is corrosive, and what steps are being taken to deal with the problem of lead contamination. Water mains containing lead pipes can be replaced, as well as those lead service connections that are the responsibility of the supplier.

• Calcium orthophosphate may be added to reservoirs to prevent lead from leaching into water.

• Well water can be treated to make it less corrosive. Calcite filters control corrosion and can be installed between the water source and any lead service connections or lead-soldered pipes.

• Reverse osmosis devices and distillation units

may be installed at the faucet to reduce the amount of lead in the water. These units can be expensive, and their effectiveness varies.

• Purchase bottled water for home and office consumption, if public water supplies are suspect. Bottled water is regulated by the FDA or by the State in which it is purchased, but controls are often fewer for bottled water than for public water supplies.

• Instruct plumbers to use only lead-free materials for repairs in the home or for new utilities. Solders and flux are considered "lead-free" when they contain not more than 0.2 percent lead. (In the past, solder contained up to 50 percent lead). Pipes and pipe fittings are considered "lead-free" when they contain not more than 8 percent lead.

• In a newly built home, remove all strainers from faucets and flush the water for at least 15 minutes to remove loose lead solder or flux debris from the plumbing.

• In homes where lead is a problem, water softeners should not be connected to pipes leading to taps supplying drinking water. A water softener may also contribute to the corrosiveness of the water and increase the potential for lead contamination.

• Note that carbon filters, sand filters, and cartridge filters do not remove all lead from water and do not prevent corrosion.

SCREENING FOR LEAD EXPOSURE

The goal of a lead poisoning screening program is to identify children with potentially toxic blood lead levels and to quickly recommend methods to reduce the environmental exposure and the level of lead in the blood. The Centers for Disease Control (CDC) guidelines of Oct, 1991 defined the potentially toxic blood lead level as 10 mcg/dL or greater. Capillary, finger prick, specimens are more practical than venous samples for screening purposes but they may be contaminated by lead containing dust on the child's hand. Diagnostic blood levels must be measured on blood obtained from a vein.

All children are at risk for lead poisoning in our industrialized society and ideally, all children should be screened. In many States, physicians and health care providers are required by law to screen children 6 months to 6 years of age for lead poisoning, in accordance with guidelines and criteria set forth in 1991 by the CDC and the American Academy of Pediatrics. A questionnaire concerning housing and other factors is used to determine which children are at high risk for lead poisoning. Physicians are required to screen children, using a blood lead measurement, if the answer to one or more of the following risk factors is "yes":

Does your child.-

1) Live in or regularly visit a house with peeling or chipping paint built before 1978?

2) Live in or regularly visit a house built before 1978 with planned or ongoing renovation or remodelling?

3) Have a brother or sister, a housemate, or a playmate with confirmed lead poisoning?

4) Live with an adult whose job or hobby involves exposure to lead?

5). Live near an active lead smelter, battery recycling plant, or other industry likely to release lead?

The validity of this questionnaire has been tested in urban and suburban areas of the United States, and the questions about the home environment were more sensitive indicators of elevated lead levels than other standard high-risk questions. An abbreviated version may be as effective as the complete questionnaire, if emphasis is given to whether the child *ever* lived in a home built before 1978, or more significantly, before 1960. Useful risk assessment questions may vary in different areas and populations.

Based on this questionnaire assessment, children are categorized as low or high risk for lead poisoning. If the answers to all questions are consistently negative, the child is at low risk for lead poisoning, and the CDC recommends a blood lead screening test at 12 months of age and at 24 months. If the answer to any question is positive, the child is considered at high risk, and a blood lead test should be obtained at the time of that examination.

Universal screening is not mandated in every state, but by 1991, 16 states had lead statutes with screening provisions, and an additional 19 had lead statutes with no screening provisions.

In Illinois, since January 1, 1993 and beginning with the 1994-95 school year, the parent of a child, 6 months - 6 years old, is required to provide a statement that the child has been screened for lead poisoning within one year prior to admission to a day care center, day care home, preschool, nursery school, or other child-care facility, licensed or approved by the State of Illinois. Similar statements are required before entering kindergarten or first grade, the fifth and ninth grades, or by age 5, 10, and 15 years, and with subsequent health examinations.

In practice, universal biannual screening and retesting guidelines are not being followed at regular intervals. Many physicians consider this form of universal screening to be impractical and not cost effective. Negative consequences to the family include, in addition to the cost of screening and of retesting, the trauma of an invasive procedure, false positives, and parental anxiety.

Physicians opposed to universal screening would advocate selective screening limited to high risk children, and those with symptoms that suggest lead poisoning. These include **loss of appetite, abdominal cramps, constipation, anemia, apathy, lethargy, periodic vomiting, clumsiness, hyperirritability, delayed development, or failure to thrive.**

Lead poisoning is a reportable disease. Physicians, nurses, laboratories, and hospitals are required to report cases of elevated blood lead levels >15 mcg/dL. Note that laboratories have a margin of error up to 5 mcg/dL either way. This variability in testing should be considered when interpreting results. Serial measurements of blood lead are more meaningful than a single test result.

COMMUNITY AND PUBLIC HEALTH ROLES

Public health officials work in collaboration with physicians and education, social service and housing agencies that have a role in community efforts to prevent lead exposure and poisoning. Lead poisoning requires a coordinated approach. Patients are diagnosed by doctors and nurses providing routine health care. Local health departments visit high-risk neighborhoods to set up screening clinics. State laboratories arrange for testing of blood samples.

State Departments of Public Health have a Case Registry for reported cases of lead poisoning in children younger than 16 years. A surveillance system collects data on sources of lead and abatement measures, treatment, and patients, which helps in formulating a public policy regarding lead poisoning in a community.

Communication between local, state, and federal agencies should deal with health, housing, environmental, and children's welfare issues and lead to joint prevention strategies. Community intervention requires to address the following to be successful:

1) Determination of populations at risk for lead exposure.

2) Education of parents, day care providers, and landlords about lead poisoning risks and preventive measures. Counselling and education of parents regarding ways to avoid lead exposure and the importance of nutrition and hygiene have been found effective in the management of children with lower levels of lead toxicity.

3) Arranging for a program of risk abatement in the community and particularly in old, high-risk

housing neighborhoods.

Guidelines for the Detection and Management of Lead Poisoning, provided by the Illinois Department of Public Health and other State Agencies for use by Physicians and Health Care Providers, should do much to prevent and provide optimal treatment for lead poisoning in children. Lead affects 17 percent of all American preschool children and approximately 175,000 young children in the State of Illinois.

The morbidity and mortality from acute encephalopathy due to high levels of lead absorption have been reduced remarkably by governmental regulations banning the use of lead in paint and lead-containing solder and pipes in plumbing. The prognosis of acute lead poisoning has been benefited by the screening programs which facilitate early detection and more successful treatment. Efforts are now being directed to the elimination of exposures to lower doses of lead which are known to result in long-term subtle but none the less, serious health problems in young children.

Baghurst PA et al. Environmental exposure to lead and children's intelligence at the age of seven years. N Engl J Med Oct 29, 1992;327:1279-84.

Baron ME, Boyle RM. Are pediatricians ready for the new guidelines on lead poisoning? Pediatrics February 1994;93:178-182.

Binns HJ et al. Is there lead in the suburbs? Risk assessment in Chicago Suburban pediatric practices. Pediatrics February 1994;93:164-171.

Campbell JR, McConnochie KM, Weitzman M. Lead screening among high-risk urban children Are the 1991 Centers for Disease Control and Prevention Guidelines feasible? Arch Pediatr Adolsec Med July 1994;148:688-693.

Centers for Disease Control, Morbidity and Mortality Weekly Report. Fatal pediatric poisoning from lead paint - Wisconsin, 1990. JAMA April 24, 1991;265:2050.

Centers for Disease Control and Prevention. Preventing Lead Poisoning in Young Children. Atlanta, GA, Department of Health and Human Services, Public Health Service, October 1991.

Division of Environmental Health, Centers for Disease Control and Prevention. The decline in blood lead levels in the United States. The National Health and Nutrition Examination Surveys (NHANES). JAMA July 27, 1994;272:284-291.

Environmental Protection Agency. Lead and Your Drinking Water. Washington, D.C., U.S. EPA, Office of Water, April 1987.

Environmental Protection Agency. Strategy for Reducing Lead Exposures. Washington, D.C., U.S. EPA, February 1991.

Illinois Department of Public Health. Guidelines for the Detection and Management of Lead Poisoning for Physicians and Health Care Providers. Springfield, State of Illinois, April 1992.

Kimbrough RD, LeVois M, Webb DR. Management of children with slightly elevated blood lead levels. Pediatrics February 1994;93:188-191.

Millichap JG et al. Lead paint: a hazard to children. Lancet Aug 23, 1952;2:360.

Millichap JG. Lead poisoning. In Environmental Poisons in Our Food. Chicago, PNB Publishers, 1993.

Millichap JG. Ed. Low level lead exposures and IQ. In Progress in Pediatric Neurology II, Chicago, PNB Publishers, 1994.

Schaffer SJ, Szilagyi PG, Weitzman M. Lead poisoning risk determination in an urban population through the use of a standardized questionnaire. Pediatrics February 1994;93:159-163.

Shannon MW, Graef JW. Lead intoxication in infancy. Pediatrics Jan 1992;89:87-90.

Shukla R et al. Lead exposure and growth in the early preschool child: a follow-up report from the Cincinnati lead study. Pediatrics 1991;88:886-892.

Tejeda DM et al. Do questions about lead exposure predict elevated lead levels? Pediatrics February 1994;93:192-194.

World Health Organization. Guidelines for Drinking-Water Quality, Vol.2, Health Criteria and Other Supporting Information. Geneva, WHO, 1984.

World Health Organization. Guidelines for Drinking-Water Quality, 2nd Edition, Vol 1, Recommendations. Geneva, Office of Publications, WHO, 1993.

CHAPTER 6

NITRATE, FLUORIDE, AND OTHER INORGANIC CONSTITUENTS

O f more than 30 health-related inorganic constituents in drinking water, eight have Maximum Contaminant Levels (MCLs) issued by the Environmental Protection Agency as National Primary Drinking Water Standards in 1991. These include arsenic, asbestos, barium, cadmium, chromium, copper, fluoride, lead, mercury, nitrate, and selenium. In 1993, five additional inorganic chemicals had MCLs applied: antimony, beryllium, cyanide, nickel, and thallium.

Only nitrate poses an immediate threat to health whenever the MCL is exceeded, but lead, fluoride and other inorganic contaminants may cause chronic health effects if exposure to elevated levels is prolonged. The World Health Organization has issued health-based guideline values for the inorganic constituents of drinking water, when these are considered appropriate.

The MCLs issued by the EPA in the United States and the WHO guideline values are compared in Table 6-1.

TABLE 6-1. EPA National Primary Drinking Water Standards and WHO Guideline Values Compared.

Inorganic Chemical	*MCL (mg/L)*	*Guideline Value (mg/L)*
Arsenic	0.05	0.01
Asbestos	7 MFL	-
Barium	2	0.7
Cadmium	0.005	0.003
Chromium	0.1	0.05
Copper	-	2
Fluoride	4	1.5
Lead	0.015	0.01
Mercury	0.002	0.001
Nitrate	10	50
Nitrite	1	3
Selenium	0.05	0.01

Adapted from U.S. EPA Report, 1991, and WHO Guidelines for Drinking Water Quality, 1993.

NITRATE AND NITRITE

Nitrates are present in soil, surface and ground waters, and in vegetables and other plants. Nitrites are formed by incomplete bacterial oxidation of organic nitrogen. Nitrates occur as fertilizers, in explosives, and as food preservatives. Nitrites are also used in food preservatives, as a coloring agent, and a corrosion inhibitor. Nitrates and nitrites are ubiquitous in the environment, occurring in foods, the atmosphere, and in most water sources.

Water Sources. Fertilized land, decayed vegetable matter, effluents, industrial discharges, leachates from refuse dumps, and atmospheric washout all may contribute to contamination of water supplies with nitrate. Well water is particularly vulnerable. Seasonal variations in nitrate concentrations may occur in rivers, and high levels are experienced after heavy rainfall.

Levels of nitrate in water are usually below 5 mg/L, but levels exceeding the MCL (10 mg/L) can result after fertilizer use in farming communities. The nitrate levels in water from the tap are similar to the source waters, since water treatment and disinfection do not modify the nitrate concentration. In contrast, nitrite levels in tap-water are considerably lower than source waters because of oxidation during chlorination.

Metabolism. Nitrate absorbed into the blood stream through the upper intestine becomes concentrated in the saliva. In the mouth and stomach, salivary nitrate is converted into nitrite by bacterial reduction. The conversion to nitrite is delayed by the acidity of gastric juice of adults but in infants, whose stomach acidity is low, nitrite formation may be higher.

Nitrites pose two important health risks: 1) Formation of methemoglobin by oxidation of hemoglobin, and 2) Formation of nitrosamines by reaction with amines and amides derived from food.

METHEMOGLOBINEMIA.

An abnormal increase in the concentration of methemoglobin in the blood occurs whenever the balance of oxidation and reduction of heme iron in hemoglobin is disturbed. Normally, 1 to 3 percent of hemoglobin is oxidized to methemoglobin each day. When methemoglobin accumulates in concentrations above 10 percent, the blood assumes a chocolate brown color and the patient's skin becomes gray blue or cyanotic. Because methemoglobin cannot bind and carry oxygen from the lungs, the patient suffers from hypoxia (lack of oxygen) of the tissues. Concentrations of methemoglobin greater than 70 percent may be fatal.

Newborns are particularly susceptible because: 1) their hemoglobin is more readily oxidized to methemoglobin than that of adults, and 2) enzymes (methemoglobin reductase) necessary to convert the methemoglobin to hemoglobin are deficient until the infant reaches 6 months of age. Methemoglobinemia may also occur in children but is rare in adults.

A variety of drugs and chemicals may increase the rate of oxidation of hemoglobin to form methemoglobin. These include acetophenetidin (Phenacetin), aniline based crayons, sulfonamides, and nitrites. Surprisingly, not a single case has been reported with ingestion of acetaminophen (Tylenol) alone, despite its close chemical relationship to acetophenetidin and aniline. In combination with other aniline drugs, however, acetaminophen may be suspect. A chronic, congenital inherited form of methemoglobinemia is caused by an enzyme (reductase) deficiency.

The most common cause of infantile methemoglobinemia is an excessive concentration of nitrate in well water used to reconstitute baby formula. Water supplies containing high levels of nitrate (>100 mg/L) should be avoided and bottled water substituted for infants. Water should not be boiled since evaporation will result in increased concentrations of nitrate.

An outbreak of methemoglobinemia in New Jersey in 1992 involved more than 40 elementary school children in first through fourth grades. They visited the school nurse within a 45-minute interval following the school lunch period. They all suffered from an acute onset of blue lips and hands, vomiting, and headache. Fourteen were hospitalized, and all recovered in 36 hours. The outbreak due to nitrite poisoning was traced to soup contaminated by nitrites in a boiler additive. The authors of the article aptly named the soup "boilerbaisse."

An adult woman secretary, aged 34 years, collapsed shortly after drinking a beverage made with water that had been accidentally contaminated by an anti-corrosive agent containing sodium nitrite. She experienced nausea, vomiting, dizziness, weakness, and diarrhea. She was cyanosed and the methemoglobin concentration was 49 percent. She recovered promptly after the intravenous administration of methylene blue.

Paradoxically, a disorder that may be caused by an aniline blue dye or drug turns the patient blue and is treated successfully with another blue aniline derivative, methylene blue. Methylene blue activates a methemoglobin reductase enzyme that facilitates conversion to hemoglobin. My colleagues at the Mayo Clinic have written a colorful account of Blue Gods (Krishna), Blue Oil (aniline/indigo), and Blue People (methemoglobinemia), with reference to the British indigo trade of the late 19th century in India. They report two cases of methemoglobinemia resulting from an anesthetic benzocaine spray.

CARCINOGENIC NITROSAMINES.
Humans may be susceptible to the carcinogenic toxicity of nitrates and nitrosamines formed in the stomach, particularly in persons with low stomach acidity. Significantly higher levels of nitrate have been found in well waters of regions in China, Columbia, and in two English towns where the incidence of stomach cancer is unusually high. These epidemiological studies are suggestive but not yet proof of a link between nitrate in water and cancer.

Nitrate consumed in meat, vegetables, or drinking water is reduced to nitrite by bacteria in the intestine. Nitrites act on amines in the diet to form nitrosamines which have been shown to cause cancer in animals. The cancer-inducing health risk of nitrates in drinking water and certain foods is a potential hazard but further investigation is ongoing.

FLUORIDE

Fluoride salts are found in fluorinated water, dental hygiene products, and dietary supplements. Drinking water accounts for 0.5 to 1.5 mg/day of fluoride, the largest single source of dietary intake.

Except for nursery brand water, with fluoride added, most bottled waters have no fluoride. Foods particularly rich in fluoride are fish, vegetables and tea. Some fish contain as much as 100 mg/kg, and tea may contain more than twice that amount.

The National Drinking Water Standard MCL for fluoride is 4 mg/L. This is high, and the Guideline value set by the World Health Organization in 1984 was 1.5 mg/L. Concentrations above this value carry an increasing risk of dental mottling and fluorosis. A safe level in relation to fluoridation of water supplies is 1 mg/L.

In acute toxic ingestion, fluoride acts as a cellular poison. It inhibits enzyme systems concerned with sugar metabolism, depresses tissue respiration, and binds with calcium. A sharp decline in the serum concentration of calcium occurs within 10 minutes of ingesting a toxic dose, causing heart irregularities and, with fatal amounts approximating 30-60 mg/kg, resulting in cardiovascular collapse. Acute hemorrhagic gastroenteritis, toxic kidney disease, and injury to the liver and heart muscle may also occur with the ingestion of a single large dose of about 2 grams. Initial symptoms and signs of intoxication are nausea, vomiting, abdominal pain, diarrhea, weakness, tetanic spasms, and sometimes convulsions.

Chronic intoxication (fluorosis) characterized by dental mottling and skeletal involvement may occur from drinking water with high fluoride concentration in excess of 4-6 mg/L over prolonged periods. Symptoms include weight loss, brittleness of bones, anemia, weakness, stiffness of joints and muscles, and calcification of ligaments and tendons.

The amount recommended for fluoridation of water supplies is 1 ppm. The incidence of caries decreases as the concentration of fluoride in water increases to about 1 mg/L. Some authorities report mottling of dental enamel and early signs of fluorosis with continuous intake of water containing more than 2.5 ppm fluoride.

Significant toxicity is unlikely from the judicious use of toothpastes which contain a maximum of 120 mg fluoride per tube. However, parents should instruct children to use small amounts of paste and to brush without swallowing the paste if possible. Mouthwashes containing fluoride should be avoided in young children.

Treatment of fluorosis includes avoidance of the source of excess fluoride, calcium supplements, and fluids. Removal from exposure for a year or more may be necessary before joint stiffness is reversed.

For acute poisoning with fluoride, treatment varies with the amount ingested. For less than 5 mg/kg, milk to relieve gastrointestinal symptoms is sufficient. For larger toxic doses, antacids or calcium chloride are given to bind or precipitate the fluoride; gastric lavage is recommended to remove stomach contents; calcium gluconate may be required to correct the low serum calcium; and intravenous fluids to induce passage of an alkaline urine. The time to recovery may be prolonged but the prognosis is usually good, if the patient survives more than 24 hours after acute poisoning.

Claims that cancer, Down's syndrome and other birth defects may be associated with elevated levels of fluoride in drinking water have not been proven, but a question of sensitivity or idiosyncratic response to fluoride in some patients cannot be dismissed.

ARSENIC

Arsenic is ubiquitous in the environment and is found in high concentrations in some well water. Water supplies in general contain very low levels of arsenic, but gross contamination of wells has occurred in special situations (e.g. in India), resulting in several thousand micrograms of arsenic per liter of water.

In China, skin cancer has occurred in association with drinking well-water containing 0.5 mg/L arsenic. The WHO has estimated that a life-time exposure to arsenic in drinking water at a concentration of 0.2 mg/L would result in a 5 percent risk of developing cancer of the skin.

Poisoning may occur with doses as low as 3-6 mg/day over extended periods. As little as 0.6 mg arsenic per liter of contaminated water was responsible for some infant deaths in one report from Chile.

Symptoms and Signs. *Acute poisoning* is manifested by jaundice, enlargement of the liver, lung congestion, kidney failure, psychosis, and encephalopathy. The patient has headaches, confusion, hallucinations, seizures, and loss of vision. After the first week, nerve involvement results in a burning sensation in a stocking and glove distribution of the legs and arms. This is followed by a generalized muscle weakness.

Chronic poisoning is characterized by skin and nail lesions, notably pigmentation (arsenic melanosis), thickening of the palms and soles (hyperkeratosis), and transverse white bands in the nails (Aldrich-Mees lines). Depigmentation of the skin and hair loss may also occur. Skin cancers appear in 5 to 10 percent of patients after latent periods of 5 to 25 years.

In diagnosis, arsenic can be measured in blood, urine, hair, or nails by absorption spectrophotometry or neutron activation techniques. A qualitative urine test (Gutzeit test) is also available.

Treatment. Dimercaprol (BAL) given within 24 hours of acute exposure is the treatment of choice. The nerve damage and kidney failure are sometimes resistant to treatment.

Prevention. Well water should be examined for possible contamination, and other sources of arsenic such as herbal preparations should be avoided. Physicians and consumers should be alert to the early manifestations of arsenic poisoning so that treatment may be prescribed promptly.

ASBESTOS

Asbestos is commonly found in water supplies, resulting from the use of asbestos-cement pipes in distribution systems. Ordinary sand filtration should remove about 90 percent of asbestos fibers. The EPA National Drinking Water Standard for asbestos is 7 million fibers per liter. The WHO in 1993 found it unnecessary to issue a health-based guideline value for asbestos because the concentrations normally found in drinking water are not considered hazardous to human health.

Asbestos is a known human carcinogen when inhaled. Epidemiological studies of populations exposed to high concentrations of asbestos in drinking water have failed to show convincing evidence of a link with cancer.

BARIUM

The EPA 1991 Drinking Water Standard for barium is an MCL of 2 mg/L, and the WHO 1993 guideline value is 0.7 mg/L. The estimated daily intake for Americans is 1 mg, mainly from food. Brazil nuts are exceptionally rich in barium.

Ingestion of large amounts of barium causes abdominal pain, vomiting, and diarrhea. Serum potassium is lowered, causing muscle weakness and heart irregularities. Studies have shown a possible association between mortality from cardiovascular disease and a 10 mg/L barium content of drinking water in some communities. Epidemics of "Pa Ping" disease in China were attributed to the prolonged ingestion of table salt containing up to 250 g/kg of barium chloride.

CADMIUM

Surface water supplies may be contaminated with cadmium by discharges of industrial wastes, leaching from landfills, and sewage polluted soils. Unpolluted waters contain less than 1 mcg/L. Higher levels of cadmium in tap-water are caused by corrosion of plumbing fittings, silver-based solders, and galvanized iron piping. The preparation and storage of food and water in galvanized containers carry a risk of cadmium poisoning.

The MCL for cadmium in drinking water is 0.005 mg/L, and the WHO guideline value is 0.003 mg/L. The estimated total daily intake of cadmium from food for Americans is about 50 mcg.

Metabolism. The absorption of cadmium from the intestine is very low, but once accumulated in the tissues, it remains and excretion is minimal. Deposited chiefly in the kidney, one third of the cadmium in the adult body has been absorbed in the first few years of life. To avoid kidney damage, it is important to prevent exposure to cadmium in infancy and early childhood.

Interactions with proteins, vitamins, and minerals affect the toxicity of cadmium. A low-protein intake and calcium and vitamin D deficiences exacerbate the adverse effects of cadmium on growth and may induce osteomalacia and spinal curvature. Pyridoxine supplements may predispose to cadmium toxicity, whereas vitamin C has a protective effect.

Symptoms of Poisoning. Acute ingestion of 5 gm, sufficient to commit suicide, caused damage to liver and kidneys and heart irregularities. Smaller amounts cause transient diarrhea.

Chronic poisoning induces kidney damage, protein in the urine, severe pains in the back and joints, waddling gait, osteomalacia, bone deformities and fractures. A disease named *Itai-itai (ouch-ouch)* caused by cadmium poisoning affected multiparous Japanese women, ages 40 to 70. Those affected lived in industrial mining areas where cadmium containing waste in river water was used to irrigate rice and for household consumption.

Prevention. There is no specific treatment for cadmium poisoning, and prevention by avoidance of exposure is all important. A diet high in protein and supplemented with vitamin C and iron may decrease the occupational risk of toxicity in workers or in consumers exposed to high levels of cadmium in water.

CHROMIUM

The estimated safe and adequate daily dietary intake of chromium varies from 10-60 mcg for infants, 20-120 mcg in young children, and 50-200 mcg in adolescents and adults. Drinking water normally contains very low concentrations of chromium (2 mcg/L or less). The levels in tap-water vary with the time of day and the length of time in contact with plumbing fittings. Contamination has occurred from industrial effluents discharged into rivers and surface waters. The MCL drinking water standard for chromium is 0.1 mg/L, and the WHO provisional guideline for a safe level is 0.05 mg/L. The daily intake in water is <10-40 mcg.

Chromium is important in the prevention of diabetes and atherosclerosis. An impaired glucose tolerance in malnourished children and in some diabetics will respond to chromium supplements. Chromium is contained in some high energy food supplements favored by body building enthusiasts. While supplements of 150 mcg/day are considered nontoxic, a varied diet is the best assurance of an adequate and safe chromium intake.

Chromium, in the hexavalent form found in drinking water, has been implicated as a factor in the cause of intestinal cancers, but only with high doses. Chronic toxicity from occupational exposure to dusts containing chromate causes an increased incidence of lung cancer. There appears to be no evidence that at current levels of non-occupational exposure to chromium a health hazard exists, according to an international cancer research agency report.

COPPER

Copper levels in drinking water generally vary from 0.01 to 0.5 mg/L, but concentrations may be higher when water stagnates in copper plumbing. Copper in water at levels above 5 mg/L may be detected by the color and an undesirable, astringent taste. Staining of laundry and plumbing fixtures occurs when copper concentrations in drinking water exceed the WHO guideline levels of 1.0 mg/L (1984) and a 2 mg/L provisional level established in 1993. Corrosion of aluminum and zinc cooking pans and galvanized iron plumbing fittings may be increased by copper. Poisoning has resulted from hemodialysis, when copper is a component of the dialysis tubing and water with an acid pH is supplied through copper plumbing.

Symptoms of Poisoning. Copper is a red blood cell poison, damaging the cell membranes and causing hemolytic anemia and jaundice. Other signs of copper toxicity include nausea, vomiting, diarrhea, abdominal pain, pancreatitis, and severe muscle pain. Acute gastric irritation can occur with drinking water concentrations above 3 mg/L.

Wilson's disease, an inborn error of copper metabolism resulting in accumulations of copper and degeneration of the brain and liver, is exacerbated by increased copper intake in the diet and water. Copper in drinking water may be a causative factor in liver cirrhosis in bottle-fed infants.

MERCURY

Mercury in inorganic form is generally present in surface and ground waters at concentrations less than 0.5 mcg/L. The EPA drinking water standard MCL for mercury is 0.002 mg/L, and the WHO total mercury guideline value is 0.001 mg/L. Pregnant women, nursing mothers, and infants are likely to be at greater risk from the adverse effects of excess mercury than the general population.

Mercury discharged into waterways and inland lakes from pulp and paper mills is converted into methylmercury by microorganisms in the water and in the digestive tracts of animals. Methylmercury accumulates in fish and fish-eating birds and animals. At each step of the food chain there is a bioaccumulation of mercury. The amount of mercury found in fish may be 3000 times the original concentration in the contaminated water. Since methylmercury is 1000 times more soluble in fats than in water, it concentrates in muscle and brain tissue and favors bioaccumulation.

An example of a large scale outbreak of methylmercury poisoning occurred in Japan in the 1950s when fishermen and their families at Minamata Bay were stricken with a mysterious neurological disease. The source of the poisoning was the consumption of fish and shellfish contaminated with methylmercury derived from materials discharged into the bay from vinyl chloride and acetaldehyde manufacturing plants. Minamata disease is symbolic of the tragic health risks of industrial pollution of drinking water sources.

Fish contaminated with mercury from industrial wastes and agricultural insecticides has become a source of concern in the Midwest Inland Lakes of the United States. Recent tests of lake water by the EPA were positive for mercury in 90 percent of samples from 380 different sources in Michigan, Illinois, Indiana, and Wisconsin. State Departments of Natural Resources publish health guides for consumers of sport fish and issue warnings of contaminated rivers and lakes. Environmental groups have called for more stringent industrial emission standards and regular consumer advisories.

Symptoms of Poisoning. With mild exposure the symptoms are subtle and diagnosis is difficult. Insomnia, nervousness, tremor, impaired judgement, fatigue, loss of sexual drive, and depression are symptoms often mistakenly ascribed to psychological causes. As this *micromercurialism* increases, the patient develops a metallic taste, abdominal cramps, diarrhea, and skin rash.

More advanced stages of chronic mercury poisoning are manifested by a wide array of serious neurological problems, including persistent tremors of the extremities, tongue, and lips; a progressive unsteadiness of gait and slurred speech; delusions and hallucinations; and inflammation of the nerves of the extremities associated with loss of sensation, numbness, and pain in the hands and feet. The term *"acrodynia"* was coined for these painful extremities in infants with chronic mercury poisoning. Acrodynia was traced in

1953 to the use of mercury teething powders, and the diseases is now rare, occurring only after accidental mercury exposure.

Treatment and Prevention. Dimercapto-succinic acid, BAL or L-Penicillamine are effective antidotes for mercury poisoning. The natural course of the disease, acrodynia, in infants and children is prolonged, with varying grades of severity. Prevention is paramount because brain damage resulting from mercury poisoning is irreversible. Public health considerations include the following preventive measures:

• Ban disposal of industrial mercurial wastes in waterways.

• Test inland lakes and other fisheries for mercury and especially methylmercury, and issue timely warnings and fishing regulations.

• Reduce the mercury content of poultry and seafoods, which account for nearly all the mercury intake of Americans.

• Eat lake fish sparingly, and avoid large trout caught in Lake Michigan.

• Avoid mercury exposure from agricultural chemicals, occupational sources, mercury-containing latex paint, dental amalgams and offices, medicines, thermometers, and household products.

SELENIUM

The level of selenium in tap-water samples from various public water supply systems rarely exceed 0.01 mg/L. Higher levels may be found in some well waters. The MCL is 0.05 mg/L, and the WHO guideline value is

0.01 mg/L of drinking water. Food is the main source of selenium exposure, particularly grains, meat, and seafood.

An epidemic of chronic selenium intoxication occurred in China in the 1960s which was attributed to selenium-laden vegetables. Symptoms of poisoning include nausea, vomiting, abdominal pain, diarrhea, joint pains, hair loss, garlic breath, bronze skin, liver dysfunction, and skin rash.

An unproven theory that increased selenium intake may modify or prevent human cancer, particularly in the intestinal tract, has promoted the sale of nonprescription selenium supplements, which may lead to an increasing frequency of chronic selenium toxicity.

SODIUM

The highest freshwater sodium concentrations are found in lowland rivers and in ground water. Upland streams and reservoirs will tend to have a low sodium content. Seawater contains about 10 g/L of sodium, and ground waters contaminated by sea waters will have increased sodium levels. Windborne sea spray and discharge of sewage effluents in rivers can add significantly to the sodium content of water supplies.

The majority of water sources contain less than 20 mg/L, but domestic water softeners and water-treatment chemicals can raise the level to 30 mg/L in some communities. The daily intake of sodium from drinking tap-water is less than 50 mg. With a relatively low sodium diet containing 2000 mg daily and a water sodium concentration of 20 mg/L, the daily intake of sodium provided by drinking water is estimated at 40 mg, which amounts to 2 percent of the total. Water would account for a larger percentage of the total sodium intake if the diet were restricted to 500 mg per day.

In areas where levels of sodium in water exceed 20 mg/L, patients with high blood pressure and congestive heart failure need to restrict their overall dietary intake of sodium. Since dietary modifications are dependent on local water conditions, no specific level of sodium in drinking water based on health considerations is recommended in the WHO guidelines.

In addition to hypertension, diseases associated with elevated blood sodium levels include hypernatremic convulsions in infants.

HYPERNATREMIC CONVULSIONS

Increased sodium levels in blood plasma can occur in infants suffering from gastrointestinal infections complicated by diarrhea and dehydration. Brain damage with convulsions are associated with hypernatremia and its treatment. The mortality is 10 - 20 percent, and one third of survivors have mental retardation and epilepsy.

Salt loading, often accompanied by excessive water loss, is one of the mechanisms of hypernatremia. Infant feeding practices using cows' milk and solid foods are contributory in some cases. Cows' milk contains three times the amount of sodium in breast milk. If tap-water of high sodium content is used to reconstitute infant formula, the tendency to hypernatremia is exacerbated. The immature infant kidney has problems in maintaining the plasma sodium at normal levels, and care must be taken to avoid excess sodium in the diet of infants.

SODIUM AND HYPERTENSION.

Epidemiological studies have shown that school children (particularly girls) living in areas with moderately raised levels of sodium in the drinking water (140 mg/L) had higher blood pressures (3-5 mm/Hg) than those exposed to low levels of sodium (28 mg/L). Girls provided with bottled water of low sodium content (8 mg/L) had lower blood pressures than those drinking only bottled water high in sodium (108 mg/L). Boys showed less consistent blood pressure differences.

The sodium content of drinking water can play a role in the etiology of elevated blood pressure, but generally the water intake is only a small contributor to the dietary sodium. In patients at risk, the use of water softeners that add to the sodium concentration should be avoided.

SULFATE

The concentration of sulfate in most water supplies is low, but levels of 20-50 mg/L are common in Eastern U.S.A., Canada, and in Europe. United Kingdom water supplies contain sulfate in the range of 4 to 300 mg/L. Sulfate is not removed from drinking water by treatment methods, and the use of the flocculant, aluminum sulfate, may add 20-50 mg/L to tap-water. Bottled mineral waters average 220 mg/L, with a range of 0 to 1200, according to one report.

Magnesium sulfate has a purgative effect at concentrations above 1000 mg/L, and lower amounts will cause diarrhea in sensitive people. A 1984 WHO guideline value of 400 mg/L, based on taste threshold concentrations for the most prevalent sulfate salts, was not included in the 1993 guidelines for drinking water quality. It was recommended that health authorities be notified of sources of drinking water that contain sulfate > 500 mg/L.

Apart from the effects on taste and the cathartic action, sulfate contributes to the corrosion of water distribution systems. An MCL for sulfate is not listed by the EPA, but a level of 250 mg/L is provided as a federal guideline under the *Secondary Drinking Water Standards.* Water with these contaminants affect the aesthetic qualities of water, such as taste, odor, and color, but generally it will not cause health problems. These qualities of water are covered in Chapter 11.

GOVERNMENTAL REGULATIONS

The Safe Drinking Water Act of 1974 gave the United States its first comprehensive national program to safeguard public drinking water. It established the national drinking water standards, which protect the health of everyone who receives their drinking water from systems serving at least 25 people or having at least 15 service connections. More than 80 percent of the U.S. population and a quarter million drinking water systems, including non-community water systems, are affected by the Act.

The amendments to the Act, introduced in 1986, issued deadlines to the EPA for specifying criteria for the filtration of surface water supplies and the disinfection of drinking water from surface and ground water souces. The amendments also increased protection of ground water, a crucial source of drinking water. They gave Indian tribes the same status as States in seeking primary responsibility for drinking water and underground injection control programs.

The major responsibility for safe drinking water rests with the water supplier, the State, and ultimately with the consumer as a concerned citizen. To stay informed about changes in standards and regulations, consult the EPA toll free Safe Drinking Water Hotline, (800) 426-4791. Consumers who are knowledgeable about the causes, symptoms and signs of waterborne diseases and intoxications may avoid some of the hazards and health risks associated with contaminated drinking water.

Askew GL et al. Boilerbaisse: An outbreak of methemoglobinemia in New Jersey in 1992. Pediatrics Sept 1994;94:381-384.

Bradberry SA, Gizzard B, Vale RA. Methemoglobinemia caused by the accidental contamination of drinking water with sodium nitrite. Clin Toxicology 1994;32:173-178.

Clarkson TW. Mercury - an element of mystery. N Engl J Med 1990;323:1137.

Dinneen SF, Mohr DN, Fairbanks VF. Methemoglobinemia from topically applied anesthetic spray. Mayo Clin Proc September 1994;69:886-888.

Lucier GW, Hook GER (eds). Cadmium. Environ Health Perspect 1979;28:1-112. *A complete issue devoted to various aspects of cadmium, including metabolism and nutritional relationships.*

Masson TJ et al. Asbestos-like fibers in Duluth water supply:

relation to cancer mortality. JAMA 1974;228:1019.

Meigs JW et al. Asbestos cement pipe and cancer in Connecticut 1955-1974. J Environ Health 1980;42:187-191.

Millichap JG. Ed. Progress in Pediatric Neurology. Chicago, PNB Publishers, 1991, pp453-4.

Millichap JG. Environmental Poisons in Our Food. Chicago. PNB Publishers, 1993.

Tuthill RW, Calabrese EJ. Drinking water sodium and blood pressure in children: a second look. Am J Public Health 1981;71:722-729.

Warkany J, Hubbard DM. Acrodynia and mercury. J Pediat 1953;42:365.

World Health Organization. Guidelines for Drinking-Water Quality. Vol 2, Health Criteria and Other Supporting Information. Geneva, WHO, 1984.

Idem. 2nd Ed. Geneva, WHO, 1993.

Wu MM, Kuo TL, Hwang YH, Chen CJ. Dose response relation between arsenic concentration in well water and mortality from cancers and vascular diseases. Am J Epidemiol 1989;130:1123. *An epidemiological study with 53 references.*

CHAPTER 7

PESTICIDES AND CANCER
RELATED POLLUTANTS

C hlorinated hydrocarbon insecticides are the most ubiquitous and persistent pesticides in the environment. As an example, traces of DDT have been recovered from dust in the atmosphere that has drifted over thousands of miles and contaminated water formed from melted snow in the antarctic. DDT is still used extensively, both in agriculture and for mosquito and malaria control, in some tropical countries. It was banned in the United States in 1972, but it persists in the environment and cannot be broken down to harmless, inactive chemicals by soil microorganisms and higher organisms. It is **nonbiodegradable.**

DDT and other chlorinated hydrocarbons are fat soluble. They concentrate in the tissues of animals and are transferred along the food chain, killing fish, birds, and mammals. DDT can bioaccumulate in fish to levels more than 10,000 times the concentration in their aquatic habitat. **Biological magnification** is the term used for this increased concentration of chemicals as they ascend the food chain from small to larger animals. The levels in human milk may exceed the legal limits permitted in cows' milk. In fact, an analysis of human milk may be the most accurate measure of the extent of contamination of our environment by pesticide residues and other toxic chemicals.

The consequences of the DDT fiasco were the partial extinction of the bald eagle and golden eagle, and interference with the reproductive behavior of the osprey, falcon, and pelican. A large number of epidemiological studies have been conducted on workers exposed to DDT, but the results were inconclusive with regard to cancer incidence and possible adverse reproductive effects. In experimental animals, however, there was evidence for carcinogenicity, and furthermore, DDT impaired reproduction in several species.

The WHO 1993 guideline value for DDT and its metabolites in drinking water was set at 2 mcg/L. It was calculated on the basis of a 10 kg child and 1 liter of water a day intake, since children may be exposed to greater amounts of chemicals in relation to their body weight than adults. This limit is recommended to protect human health, but the protection of the environment and aquatic life is not guaranteed. In some countries it is argued that the benefits of DDT in the control of malaria and disease-carrying insects far outweigh any health risk from DDT in drinking water. Food is the major source of DDT exposure and more than 90 percent of the DDT stored in man is derived from food.

SYNTHETIC ORGANIC CHEMICALS (SOCs)

Most pesticides are man-made, carbon-containing chemicals called synthetic organic chemicals or SOCs. The list of those regulated by the EPA includes 2,4-D, lindane, methoxychlor, endrin, 2,4,5-TP, and toxaphene. Health effects associated with SOCs include disorders of the nervous system, liver, and kidney, and some are potentially carcinogenic.

Lindane. Contamination of water has occurred from its direct application for the control of mosquitoes, its use in agriculture, and its occurrence in wastewater from manufacturing plants. It is a frequent contaminant of surface waters in minute amounts. The MCL is 0.0002 mg/L.

Toxic Symptoms. Poisoning causes central nervous system effects, with irritability, unsteadiness, and convulsions. Treatment of scabies with lindane has resulted in headache, nausea, vomiting, spasms, and weak respirations. Occupational exposure for prolonged periods has resulted in toxic liver damage. The health risk of lindane in drinking water is low, and the WHO guideline value is 2 mcg/L.

Chlordane. Chlordane is classified as a polycyclic chlorinated insecticide. Pesticides with similar chemical structure include heptachlor, aldrin, and dieldrin. Except for termite control, the use of these insecticides was banned in the early 1980s. Household wells may become contaminated with chlordane after a house is treated for termites, but municipal water supplies are rarely affected. Drinking water is generally an insignificant source of chlordane.

Poisoning by ingestion, inhalation, or skin contamination causes hyperexcitability, tremors, unsteadiness, and convulsions, followed by depression of the central nervous system and respiratory failure. The liver function may also be impaired.

In animal studies, carcinogenicity has only been demonstrated in mice. In two clinical studies, exposure to chlordane was followed by an increased incidence of neuroblastoma and acute leukemia. Acceptable daily intakes of these chemicals and guideline values have been recommended by the WHO. (see Table 7-1).

VOLATILE ORGANIC CHEMICALS (VOCs)

Most are industrial chemicals and solvents, including degreasing agents, varnishes, and paint thinners. They are commonly referred to as VOCs. The list includes trichloroethylene, carbon tetrachloride, vinyl chloride, 1,2-dichloroethane, benzene, para-dichlorobenzene, 1,1-dichloroethylene, and 1,1,1-trichloroethane. Carcinogenic effects in animals have been demonstrated, but studies in humans are not conclusive. Acute poisoning affects all organs of the body, especially the liver.

MCLs in drinking water for 3 additional VOCs were set by the EPA in 1993. These are dichloromethane (0.005 mg/L), 1,2,4-trichlorobenzene (0.07 mg/L), and 1,1,2-trichloroethane (0.005 mg/L).

POLYCHLORINATED BIPHENYLS (PCBs)

Despite the fact that PCBs were banned in the

1970s, dangerous concentrations persist in the water of rivers and inland lakes, where they were dumped years ago as waste products from electrical transformer, capacitor, and plasticizer factories. From the sediment at the bottom of harbors, hazardous waste residues pollute the water and are eaten by microscopic organisms and fish. From fish the chemicals are passed on to birds and humans, causing various ailments including probable cancer. Man is the final consumer in the food chains, and he is exposed to the greatest concentrations of any environmental poison. The MCL for PCB contamination of drinking water is 0.0005 mg/L.

While regulatory control measure have substantially reduced the PCB contamination of animal feeds and food products by spillage, those subgroups of the population who regularly consume fish caught in lakes and streams are still at risk of poisoning. Surveys and analyses of fish from the Hudson river in New York State and from Lake Michigan showed significant levels of contamination with PCBs in excess of 5 ppm.

Historically, environmental toxicology studies in the Great Lakes basin began in the 1950s when declines in lake trout and in 14 species of wildlife were correlated with DDT and PCB contamination of lake water. Most species of fish were affected, and residue levels were proportional to the age and size of the fish. Trout and salmon had the highest levels of contamination, and whitefish had the lowest PCB residues.

Studies designed to assess the health hazard of PCB exposure from Lake Michigan fish showed a correlation between the quantity of fish consumed and the concentration of PCB in blood and breast milk of participants. Those eating higher amounts of fish had significantly higher blood levels of PCB.

In other studies, PCBs were reported in samples of malignant breast tissue. Children born to women who routinely consumed Lake Michigan sportfish displayed poorer short-term memory function on both verbal and quantitative tests in a dose-dependent fashion.

Symptoms of PCB Poisoning. The symptoms attributed to PCB human exposure include acne, skin pigmentation, eye discharge, visual impairment, weakness, numbness, headaches, and liver dysfunction. Infants born to mothers exposed to PCB were pigmented, and the skin discoloration disappeared as the child became older. The adults experienced a slow improvement in symptoms and eventual recovery. These observations were recorded in Japanese families exposed to PCB in 1968, when a heat exchanger leaked PCB into rice oil (Yusho).

DISINFECTION BY-PRODUCTS

When surface water containing organic matter is treated with chlorine, the by-products of this disinfection process are the carcinogenic chemicals, trihalomethanes. The total trihalomethanes (TTHM) include chloroform, bromoform, bromodichloromethane, and dibromochloromethane. The MCL for TTHMs in drinking water is 0.1 mg/L. This applies only to systems using chlorine disinfection and serving 10,000 or more people.

Chlorine is an effective disinfectant and is widely used to prevent outbreaks of waterborne bacterial infections. The levels of the THM by-products in water at the faucet are considerably higher than the levels in surface or ground water, in which they are generally absent. The health effects of THMs have been studied mainly in relation to chloroform, the predominant by-product of water chlorination.

Carcinogenic Effects. Animal studies have demonstrated liver cancers, and some epidemiological studies in humans showed positive correlations between rectal-intestinal and bladder cancer mortality rates and chloroform levels in the drinking water. However, the National Academy of Sciences Safe Drinking Water Committee concluded that a causal relationship between chloroform and cancer had not been established, in its review of 13 epidemiological studies. The WHO issued a guideline value for chloroform in drinking water of 30 mcg/L in 1984, and this was increased to 200 mcg/L in 1993. These concentrations would lead to one additional case of cancer per 100,000 of the population in a lifetime.

DRINKING WATER STANDARDS

The EPA sets drinking water standards for all public water systems. These are of two types: primary and secondary. Primary standards are based on health effects and are enforceable. Secondary standards are federal guidelines regarding the color, taste, and other aesthetic qualities of water and are not enforceable. These will be discussed in Chapter 11.

Primary Drinking Water Standards may be Maximum Contaminant Levels (MCL) or Treatment Technique (TT) Requirements. A MCL Goal (MCLG) is an unenforceable health goal equal to the maximum level of a contaminant not expected to cause adverse health effects over a lifetime of exposure. For chemicals which are believed to cause cancer, the MCLG may be set at zero since there is no known safe level.

The MCL is the enforceable standard, and this is set as close to the MCLG as is practical. The MCLs for some organic chemical contaminants of drinking water are listed in Table 7-1. Others were added in 1993, bringing the total number of standards to 84. The EPA goal of 111 standards for 83 substances, as mandated by Congress, should be completed by 1995.

Treatment Technique Requirements are set for those contaminants which are difficult or too costly to measure. Specific TTs include filtration or corrosion control to prevent health problems. TTs take the place of MCLs for certain chemicals and especially for bacterial, parasitic, and viral contaminants.

CANCER CAUSING CHEMICALS

Of 40 organic chemical contaminants of drinking water listed in Table 7-1, 20 have been linked to a risk of cancer in humans. The degree of cancer risk is listed as definite in 3, probable in 13, and possible in 4.

The International Agency for Research on Cancer has categorized the potential carcinogenicity of chemicals in humans in five groups:

- Group 1. Carcinogenic.
- Group 2A. Probably carcinogenic.
- Group 2B. Possibly carcinogenic.
- Group 3. Not classifiable as to carcinogenicity.
- Group 4. Probably not carcinogenic.

TABLE 7-1. National Primary Drinking Water Standards.

Contaminants	Health Effects	MCL	Sources
Organic Chemicals			
Acrylamide	cancer, nervous	TT	sewage
Alachlor	cancer	.002	herbicide
Aldicarb	nervous system	.003	insecticide
Atrazine	cardiac	.003	herbicide
Benzene	cancer	.005	fuel, solvent
Carbofuran	reproduction	.04	insecticide
Carbon tetraCl	cancer	.005	coolants
Chlordane	cancer	.002	insecticide
2,4-D	liver, kidney, cns	.07	herbicide
DiBrClpropane	cancer	.0002	fumigant
DiClbenzene	cancer	.075	moth balls
DiClbenzene o-	cns, lung, liver	.6	solvent
Dichloroethane	cancer	.005	insecticides
DiClethylene	liver, kidney	.007	dyes, paints
DiClethylene(cis)	cns, liver	.07	solvent
DiClethylene(trans)	cns, liver	.1	solvent
DiClpropane	cancer, liver	.005	fumigant
Endrin	cns, kidney	.0002	insecticide
EpiClhydrin	cancer, liver	TT	epoxy resin
Ethylbenzene	kidney, liver	.7	gasoline
Ethylene DiBromide	cancer	.00005	gasoline
Heptachlor	cancer	.0004	insecticide
Lindane	cns, liver	.0002	insecticide
Methoxychlor	cns, liver	.04	insecticide
MonoClbenzene	kidney, liver	.1	pesticide
PentaClphenol	cancer, liver	.001	herbicide
PCBs	cancer	.0005	transformer

TABLE 7-1 contd. (Adapted from U.S. EPA Report, 1991).

Organic   Chemicals

Styrene	liver, cns	.1	plastics
TetraClethylene	cancer	.005	solvent
Toluene	kidney, cns	1	gasoline
Trihalomethanes	cancer	.1	Cl water
Toxaphene	cancer	.003	insecticide
Silvex 2-4-5TP	cns, liver	.05	herbicide
TriClethane	nervous system	.2	food wraps
TriClethylene	cancer	.005	paints
Vinyl chloride	cancer	.002	pipes
Xylenes	liver, kidney	10	paint, ink

Inorganic   Chemicals

Arsenic	skin, nervous	.05	pesticide
Asbestos	cancer	7 MFL	pipes
Barium	blood	2	paint
Cadmium	kidney	.005	plumbing
Chromium	liver, skin, gi	.1	textiles
Copper	gastrointestinal	TT	plumbing
Fluoride	bones	4	additive
Lead	cns, kidney	.05	plumbing
Mercury	kidney, cns	.002	fungicide
Nitrate	"blue-baby"	10	fertilizer
Selenium	nervous system	.05	mining
Radionuclides	cancer		nuclear waste

Microbiological

Giardia lamblia	Giardiasis	fecal matter
Legionella	Legionnaires	aerosols
Coliforms	gi disease risk	fecal matter
Viruses	gastroenteritis	fecal matter

_____

MCL, mg/L; TT, treatment required; cns, nervous system.

Cancer Risk Assessment. The determination of potential cancer risk of chemicals is usually based on long-term animal studies. Data are also available on the incidence of cancer in humans, mostly from occupational exposure. Epidemiological studies of populations exposed to surface, upland, chlorinated, and reuse waters have demonstrated an increased incidence of cancer of the gastrointestinal and urinary tracts.

As one example of a population based approach, United States case-control studies in New York counties compared mortality from gastrointestinal and urinary tract cancer with noncancer death controls. There was an excess of male deaths in chlorinated surface water areas from cancer of the esophagus, stomach, large intestine, rectum, liver and kidney, pancreas, and bladder, and an excess of female deaths from cancer of the stomach.

In England, the relative risks of River Thames reused water and upland water compared to ground water showed a small but statistically increased risk of stomach cancer mortality in persons supplied by upland and river water for drinking purposes.

Water Source and Risk. Upland surface water and polluted river sources that have been chlorinated carry the highest risk of cancer. Unchlorinated ground water has the lowest cancer risk. Studies have shown sufficient evidence of a cancer risk associated with organic micropollutants in certain drinking waters that a causal relationship has been accepted. Despite the small relative risk, the high prevalence of polluted and chlorinated waters in public supplies demand the enforcement of EPA regulations to monitor the MCLs of organic pollutants.

The WHO presents guideline values, the concentration of a chemical in drinking water associated with an estimated risk of one additional case of cancer per 100,000 population over a lifetime or 70 years exposure. These values are rough estimates and usually err on the side of caution. For organic constituents, they are expressed in mcg/L of drinking water.

Consumer Assessment Factor. Consumers vary in the way they view these hazards and their respective risks. Some will accept a small risk whereas others will view any risk as unacceptable and cause for alarm. The risk imposed on us by pollutants in the public drinking water is regarded as a hazard to our health, whereas the risk we voluntarily take by drinking alcohol and coffee is of no concern.

"Is the Drinking Water Safe?" not "How Safe is Our Drinking Water?" is the question asked by many consumers at a personal level. Unfortunately, a simple "yes" or "no" answer is not possible, given the many variables in water sources, environmental factors, and treatment processes. Clearly, there is a risk involved by ingesting low levels of organic chemical contaminants every day of our lives, and that one additional case of cancer per 100,000 population might be someone close. The more we understand about the nature of the risk and the quality of our own local water supplies, the more we as consumers and citizens can do to protect ourselves from these hazardous environmental pollutants and their sources.

Chemical Pollutant Sources. Man-made cancer-related hazards in drinking water are derived from industrial, domestic, and agricultural wastes, the polluted atmosphere, leaching from soil, land runoff, motorboat engines, and spilled chemicals. The disinfection of water with chlorine is the source of cancer risk receiving most attention from the EPA. Of special concern and interest are the chlorine breakdown products, trihalomethanes (TMHs), particularly chloroform.

The FDA in 1976 restricted the use of chloroform that previously had been employed in cough syrups, mouthwashes, toothpastes, and other common consumer products. The cancer risk of exposure to chloroform is lower now than 20 years ago, but compared with other chemical pollutants, chloroform in drinking water is a significant hazard. Compared to occupational hazards and leisure-time activities, however, chlorinated water has a relatively low health and fatality risk. For example, the risk of death from cigarette smoking is estimated at 2000 times that of drinking chloroform contaminated water. Eating fish contaminated with PCBs and DDT carries a higher potential risk of cancer than drinking chlorinated water.

Association of State Drinking Water Administrators. The Regulation of Drinking Water. Arlington, VA, ASDWA, 1990.

Dreisbach RH. Handbook of Poisoning. 7th ed. Los Altos, CA, Lange Med Publ, 1971.

Goodman LS, Gilman A. The Pharmacological Basis of Therapeutics. 5th ed. New York, Macmillan, 1975.

Millichap, J.G. Environmental Poisons in Our Food. Chicago, PNB Publishers, 1993.

Ram NM, Calabrese EJ, Christman RF. (Eds). Organic Carcinogens in Drinking Water. Detection, Treatment, and Risk Assessment. New York, John Wiley, 1986.

United States Environmental Protection Agency. The Safe
     Drinking Water Act. San Francisco, EPA, September
     1990.

Idem. Is your drinking water safe? Washington, D.C., EPA,
     Office of Water, September 1991.

World Health Organization. Guidelines for Drinking-Water
     Quality. Vol 2. Health Criteria and Other Supporting
     Information. Geneva, WHO, 1984.

Idem, and United Nations Environment Programme.
     Carcinogenic, Mutagenic, and Teratogenic Marine
     Pollutants: Impact on Human Health and the
     Environment. In: Advances in Applied Biothechnology
     Series, Vol 5. Houston. Gulf Publishing Co, 1990.

Idem. Guidelines for Drinking-Water Quality. 2nd edition. Vol
     1: Recommendations. Geneva, WHO, 1993.

CHAPTER **8**

RADIATION EXPOSURE
HEALTH RISKS

R adioactive chemicals enter the environment from naturally occurring and man-made sources. Naturally occurring sources include radium-226/228 and radon-222, present in rocks and soil, and radioactive materials produced by cosmic rays. Man-made sources are the radionuclides from radioactive waste, nuclear power facilities, nuclear tests, and medical uses.

The degree of exposure to radionuclides depends on several factors:
- height above sea level,
- amount of radioactivity in the soil,
- amount released by man-made sources,
- amount absorbed from air, food, and water.

The amount contributed by drinking water to the total radioactivity exposure of a person is small, and is derived mainly from naturally occurring radionuclides in uranium deposits. Exposure to natural sources contributes more than 98 percent of the radiation dose affecting the population. Nuclear power facilities and nuclear weapon testing are generally very small contributors to the overall human exposure.

Levels of radionuclides in water are affected by man-made contaminations from nuclear facilities and medical uses. Increases in radioactive waste will add to the widespread contamination of the environment, including surface and ground water sources, which in turn leads to contamination of public water supplies. Although these sources of radioactive exposure are a relatively small percentage of the total, they have been limited in the United States by the EPA regulatory control mechanisms.

The MCLs for the various radionuclides (beta particle, gross alpha particle, and radium 226/228) are shown in Table 8-1. The WHO 1993 guidelines for acceptable screening values in drinking water are 0.1 Bq (becquerel)/L for gross alpha activity, and 1 Bq/L for gross beta activity.

Health Risks. Naturally occurring alpha activity is contributed mainly by radium 224/226, and man-made beta activity is represented by strontium-90. Exposure to ionizing radiation may result in two types of health risk: 1) occurring only at a dose above a certain threshold level, and 2) that which occurs at any dose level received over extended periods. Generally, the dose received from natural sources of radioactivity and routine exposures is well below any threshold level. Radionuclides absorbed by the body may persist, and the total dose received from all sources over months or years has potential health effects such as cancer, independent of a threshold dose level.

TABLE 8-1. EPA National Primary Drinking Water Standards for Radionuclide Contaminants, 1991.

Radionuclide	MCL	Sources
Beta particle and photon activity	4 mrem/yr	radioactive waste, uranium deposits, nuclear facilities
Gross alpha particle activity	15 pCi/L	radioactive waste, uranium deposits, geological/natural
Radium 226/228	5 pCi/L	radioactive waste, geological/natural

MCL, maximum contaminant level; mrem, 1/1000 rem; pCi, picocurie

EFFECTS OF RADIATION INJURY

Ionizing radiation produces injury in relation to the dose, and is independent of the type of particle exposure. Differences in the degree of injury are a function of quantity rather than quality of radioactivity. Effects on cells can result in death of the person exposed, death of cancer cells treated with roentgen rays, genetic mutations, and the production of cancer as a late effect of exposure to irradiation.

Adverse effects of irradiation vary not only with the dose, but also in relation to the age of the person exposed. The fetus in utero and children with rapidly differentiating tissues are more susceptible to radiation injury than adults. An infant's longer life span also allows more time for delayed effects of irradiation to develop.

An overwhelming dosage, such as an atomic bomb, will cause death or acute symptoms followed by an increased incidence of leukemia or cancer. Low-level radiation, as in persons exposed to fallout from nuclear tests in Nevada in the 1950s, was thought to induce cancer in children in Utah and in some military personnel at the test site. The level of exposure obtained from drinking water is insignificant compared to nuclear fallout and roentgen therapy, but over a lifetime, the quantity of radionuclides absorbed may be sufficient to induce cancer in susceptible persons.

Acute Symptoms. Symptoms of radiation sickness caused by doses as large as 100 roentgens are malaise, fever, nausea, vomiting, and diarrhea. These are followed by a lowering of the white blood cells, depression of the bone marrow, and loss of hair.

Delayed Effects. A significant rise in the incidence of leukemia occurred within ten years following the Hiroshima incident. Children were more susceptible than adults, having a greater incidence of leukemia and of breast cancer in later life. Those exposed in utero were born with a small head circumference, delayed growth rate, and mental retardation. In utero exposure to diagnostic radiation is reported to increase the risk of death from cancer during childhood. Ultrasound can often be used in place of CT scan or other X-ray procedures in infants.

The toxic effects of irradiation to the head, used to prevent spread of acute leukemia to the nervous system in children during treatment with chemotherapy, are well documented. Deficits in cognitive function, involving visual-spatial-motor coordination and memory, and brain atrophy have been observed in patients receiving both irradiation and chemotherapy, whereas those treated with chemotherapy alone had normal intellectual abilities after recovery.

Since radiation exposure and effects on the body are additive throughout life, it is important to limit diagnostic X-rays and therapies in childhood to those that are absolutely essential. Routine dental and chest films, bone scans requiring radioisotopes, and roentgen therapy for certain tumors should be avoided if alternative diagnostic and therapeutic methods are available.

RADON HEALTH RISKS

Radon gas exists naturally underground and is produced from the radioactive decay of uranium deposits in the soil. Its concentration varies in geographical locations. As a gas, it can enter homes in two ways. First, radon can seep through soil into cracks in household foundations. Second, it can seep through soil into well water. When well water is agitated at warm temperatures in the home, as in a shower, a washing machine, or even a faucet, radon is released into the air. Airborne radon has been linked to 13,000 deaths from lung cancer each year in the United States.

Consequently it is a water contaminant of concern to public health officials and house-holders.

The form of water supply, the various uses of water in the home, and the construction of houses are different throughout the world. Consequently, an acceptable universal drinking water level of radon has not been determined. The global average dose from inhalation of radon is estimated at 1 mSv(sievert)/year, which amounts to half the total natural radiation exposure. In comparison, the global dose derived from drinking water is relatively small. Local health departments provide information about exposure risks in their regions. Homes with radon levels greater than 4 pCi/L should be modified to reduce exposure.

Behrman RE (Ed). Nelson Textbook of Pediatrics. Philadelphia, WB Saunders, 1992.

Ciesielski KT et al. Hypoplasia of the cerebellar vermis and cognitive deficits in survivors of childhood leukemia. Arch Neurol Oct 1994;51:985-993.

Millichap JG. (Ed). Brain radiotherapy and cognition. In: Progress in Pediatric Neurology II. Chicago, PNB Publishers, 1994.

United States Environmental Protection Agency. Is Your Drinking Water Safe? Washington, D.C., Office of Water, EPA, Sept 1991.

World Health Organization. Guidelines for Drinking Water Quality, Vol 2, Health Criteria and Other Supporting Information. Geneva, WHO, 1984.

World Health Organization. Guidelines for Drinking Water Quality, 2nd edition, Vol 1: Recommendations. Geneva, WHO, 1993.

CHAPTER **9**

WELL WATER HAZARDS
AND SAFETY TIPS

Wells, cisterns, and springs account for 15 percent of our nation's drinking water supplies. These individually owned and operated sources of drinking water are licensed or registered in 46 states, and their construction standards are dictated in 42 of the states. The household well owner, however, has the final responsibility for the purity of his water supply. This chapter deals with the potential contaminants of well water, the health risks, and ways to avoid them.

Ground water contaminants are either naturally occurring or a result of human activities. Those from natural sources include nitrates and nitrites, microorganisms, radionuclides, metals (e.g. arsenic), and fluoride. Potential sources of contamination from human activity include bacteria, nitrates, lead and copper, fertilizers and pesticides, industrial products and wastes, household wastes, and water treatment chemicals. These potential well and ground water contamination sources are illustrated in Figure 9-1.

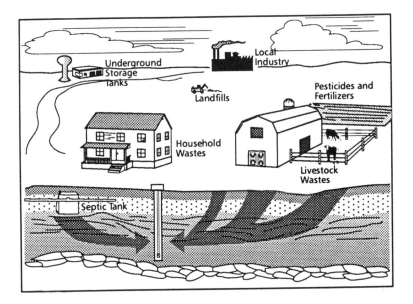

Figure 9-1. Potential Well and Ground Water Contamination Sources.

Reproduced from The U.S. Environmental Protection Agency, Office of Water (WH-550) Booklet on "Drinking Water From Household Wells," September 1990.

COLIFORM BACTERIA

These microorganisms are the most frequent contaminants found in private well water. Septic fields, due to their location, are often the source of such contamination from animal or human wastes. Shallow or improperly sealed or placed wells are chiefly affected. The degree of contamination is also dependent on the rainfall and the nature of the soil, among other factors.

Although they are not necessarily disease-causing organisms themselves, coliforms in drinking water may be indicators of contamination with additional organisms that cause gastroenteric infections, dysentery, hepatitis, typhoid fever, cholera, and other waterborne illnesses. The Environmental Protection Agency (EPA) regards the total coliform count as an indicator of fecal contamination. For small water systems, the Maximum Contaminant Level is one positive sample among 39 or fewer routine samples per month.

The International Standards Organization recommends a minimum of one sample monthly for a population of less than 5000 served by a distribution system. The EPA requires non-community water systems serving fewer than 1,000 people to collect one routine sample per quarter. Four repeat samples are required if coliforms are detected. For private household wells, the frequency of testing should reflect the local conditions.

Bacteriological surveys of ground water quality and drinking water wells in six geographical locations in the United States, between 1975 and 1980, showed much variability in the incidence of coliform contamination. Positive tests for coliforms ranged from a low of 9 percent of 621 samples taken from a community water supply to as high as 85 percent of 460 samples collected from South Carolina rural supplies.

Pathogenic microorganisms are a major cause of waterborne disease outbreaks in the United States, and ground waters obtained from deep aquifers are not exempt. Periodic checks of well water for total coliform contamination are recommended. The health risks of various waterborne microorganisms are discussed in Chapters 2, 3, and 4.

NITRATES

Nitrates are the second most common type of contaminant of private well water. The presence of nitrates suggests that animal or human wastes or fertilizers used in agriculture or on lawns are entering the well. Nitrates are of special concern to the health of infants and young children, and in women of child-bearing age. Chapter 6 covers the health risks of nitrates, arsenic, and other inorganic constituents.

LEAD

Lead infiltrates drinking water through the corrosion of plumbing materials. The sources of lead in drinking water include lead pipes, lead-based solder, and brass faucets used in household and distribution systems. The acidity or pH, temperature, and mineral content of the water affect the degree of corrosion.

The age of the plumbing materials, particularly copper pipes soldered with lead, is important. The newer the plumbing, the less protective deposits and lining on the insides of the pipes, and the greater the susceptibility to corrosion and leaching of lead.

Submersible Well Pumps. The brass and bronze alloys used in some types of submersible well pumps contain lead. Lead can leach into drinking water when the well water is in constant contact with the brass in the pump. This could create a health risk, especially if the water is soft and corrosive.

These pumps are being evaluated for certification by the National Sanitation Foundation International (NSF). Some submersible well pumps are made of stainless steel and plastic and do not pose a problem of lead contamination.

If a pump is known to contain brass or bronze components, the water should be tested for lead. The protocol for collection of the sample and delivery to a State certified laboratory for testing may be obtained from the EPA's Safe Drinking Water Hotline (800-426-4791). The maximum contaminant level (MCL) for lead in drinking water is now 15 parts per billion (ppb, or mcg/L). Levels above 15 ppb may be injurious to health, particularly in small children and during pregnancy. The health risks of lead are covered in Chapter 5.

A number of cartridge and reverse osmosis filtering units are capable of removing lead from drinking water at the tap. Various home water treatment units are discussed in Chapter 12. The Water Quality Association will advise on water filters for specific purposes (Tel. 708-505-0160). The installation of a filtering unit may be more economical than using bottled water or replacing the pump.

RADON

Radon exists naturally underground. As a gas, it can enter homes in two ways: 1) by seepage through soil into cracks in the house foundations, and 2) by seepage into well water. When well water is agitated at warm temperatures in the home, as in a shower or washing machine, radon is released into the air. Airborne radon has been invoked as a cause of lung cancer. The dose of radon exposure from all sources must be considered when some sort of intervention is required.

PESTICIDES AND FERTILIZERS

In agricultural areas, around golf courses, and in suburban households with lawns and gardens, pesticides and fertilizers are applied frequently. Chemicals used to treat termites can also pose a hazard. Many of these chemicals may seep into ground water and household wells. Fertilizers rich in nitrate may add to the risk of levels of nitrate in water that can be a potential hazard to infants and young children.

EPA National Survey. The EPA has completed a five-year National Survey of Pesticides (NPS) in drinking water wells. The Phase I results of this survey indicate that at least half of the nation's drinking water wells contain detectable amounts of nitrate, with a small percentage at concentrations higher than the EPA's regulatory and health-based limits for drinking water (1.2% of community water system (CWS) wells and 2.4% of rural domestic wells).

EPA estimates that one half of the 100,000 community water systems in the United States contain nitrate, 10 percent are contaminated with one or more pesticides, and 7 percent contain both nitrate and pesticides. Of 10 million rural domestic wells, 57 percent contain nitrate, 4 percent have pesticides, and 3 percent contain both.

Four percent of wells surveyed contained nitrate concentrations in excess of the MCL (>10 mg/L), but less than one percent of wells (approximately 60,000) and community water systems contained pesticides in concentrations exceeding the Maximum Contaminant Levels (MCL) or Lifetime Health Advisory Levels (HAL).

Nitrate is found in some fertilizers, and is also formed by nitrification of ammonium fertilizers. It is present in septic systems, animal feed wastes, industrial wastes, and sanitary landfills.

The pesticides detected most frequently in the Survey were DCPA acid metabolites and atrazine. The maximum concentrations of DCPA metabolites and atrazine detected in wells were 2.4 and 7.0 mcg/L, respectively. DCPA, also known as Dacthal, is a herbicide used primarily as a weedkiller on lawns, turf, and golf courses, and on some fruits and vegetables. Atrazine is used to control annual broadleaf weeds and grasses in corn and soybean farms. DCPA metabolites and Atrazine have a tendency to migrate through soil because of their high water solubility.

The concentrations of pesticides were usually well below levels considered as a health risk. However, five pesticides (alachlor, atrazine, dibromochloro-propane [DBCP], ethylene dibromide [EDB], and Lindane) were found in rural domestic wells at levels above their MCLs/HALs.

MCLs do not apply by law to domestic wells, but this standard was used in the EPA Survey as a measure of the quality of the drinking water. HALs are the maximum concentrations of contaminants in water that may be consumed safely over an average human lifetime.

Farming and Land Use Near Domestic Wells

The Survey questionnaire showed the following potential sources for well water contamination with

nitrate and pesticides:
- 12% located on farming properties;
- 65% located within 1/2 mile of lands farmed or used as pasture in the last 3 years;
- 5% located within 1/2 mile of a golf course;
- 94% located within 1/2 mile of a surface water body;
- 18% located within 1/2 mile of irrigated land.

About 11% of domestic wells were located in counties with high pesticide use, and 27% with high ground-water vulnerability. Farm pesticides had been used in the last five years at about 84% of farmed properties, and 40% of the pasture land with well sites. The Survey also showed that pesticides had been used in 75% of homes, 40% of lawns, and 42% of gardens connected to properties with wells. Pesticides were also stored at two thirds of the properties, with a risk of accidental spill.

Structure of Rural Domestic Wells

Most domestic wells are located on exposed or vegetated soil that is permeable to precipitation and contaminants. Septic units are usually constructed on the same property. Most wells are closed (i.e., capped) at the ground surface; 85% are cased, and 47% of cased wells are cased to their total depth. Most have not been redrilled, and 24% are more than 20 years old. Only 22% are deeper than 200 feet. The relation between well depth, age, and construction, and the causes and frequency of water contamination will be covered in a NPS Phase II Report.

The evidence of widespread migration of chemicals into wells demonstrates a need for continued attention to ground water protection. Analyses of the health risks of the various pesticide and nitrate contaminants uncovered in this extensive governmental Survey should provide important information regarding the safety of our water supplies.

INDUSTRIAL AND HOUSEHOLD WASTES

A variety of chemicals used in industry or business may become well water contaminants if their disposal is not properly controlled and supervised. Spills and wastes from gas stations and dry cleaners may threaten ground water sources tapped by wells.

Petroleum products and chemicals stored in underground tanks may leak into ground water and contaminate wells. Waste materials from landfills and dumps may seep into surrounding land, carrying various noxious chemicals at times of heavy rainfall or flooding.

Careless disposal of cleaning fluids, paints and thinners, soaps and detergents, and other household wastes may endanger the safety of a domestic well. A septic system incorrectly located in close proximity to a well will pose a potential risk of water contamination. The improper use or storage of water treatment chemicals, including chlorine and corrosion inhibitors, might contaminate the wellhead.

WELL OWNER SAFETY TIPS

The following precautions should be observed to maintain the safety of the drinking water obtained from a domestic well:

• Avoid potential sources of contamination;

• Have the well water tested periodically for nitrate and coliform organisms (also arsenic in some areas);

• Have the test results interpreted by a certified laboratory;

• Maintain the well regularly, and keep records of test results and treatments;

• Consult a local expert (e.g., health department sanitarian) if a sudden change in the appearance, taste, or smell of the water is noted and contamination is suspected;

• Call EPAs Safe Drinking Water Hotline (800-426-4791) for the name and telephone number of the State certification officer;

• Call the National Water Well Association (614-761-1711) for names of certified water well contractors in the area;

• Consult the Water Quality Association (708-505-0160) and/or the National Sanitation Foundation (313-769-8010) if information concerning an appropriate water treatment device is required.

Millichap JG. Environmental Poisons in Our Food. Chicago, PNB Publishers, 1993.

United States Environmental Protection Agency. Drinking Water From Household Wells. Washington, D.C., Office of Water, EPA, Sept 1990.

Idem. National Survey of Pesticides in Drinking Water Wells. Phase 1 Report. Washington, D.C., Office of Water, Office of Pesticides and Toxic Substances, EPA, Nov 1990.

Idem. Environmental Fact Sheet. Lead Leaching from Submersible Well Pumps. Washington, D.C., EPA, 1994.

World Health Organization. Guidelines for Drinking-Water Quality. Vol 1: Recommendations, 2nd ed, Geneva, 1993.

CHAPTER **10**

BOTTLED WATER USAGE
AND SAFEGUARDS

B ottled water is a luxury to many and a necessity to others. Company advertising has promoted bottled water as "purer," "safer," "chlorine-free," and an essential component of a healthy life-style. With a growing concern over reports of hazardous contaminants in public water supplies, Americans are resorting to bottled water, hoping to avoid the health risks that sometimes attend the use of water from the tap.

According to an International Bottled Water Association (IBWA) survey, Americans consumed more than 1.5 billion gallons of bottled water in the year 1989, close to seven gallons per capita. By the year 2000, the IBWA projects an increased usage of bottled water to 19 gallons per capita. More than 600 different brands of bottled water are produced in the United States, and over 50 brands are imported from other countries, especially Europe where bottled water is the customary form of drinking water.

LUXURY USAGE

Some prefer bottled to ordinary tap water only because of the taste. According to consumer attitude and usage surveys conducted by the IBWA, taste is the major reason why consumers buy bottled water. Ozone disinfection is used by the majority of bottling facilities, and unlike the chlorine used to disinfect tap water, ozone leaves no chemical residual after taste or smell.

Others find it convenient to carry while exercising on a bicycle or at a health club, or at conferences or workshops, and preferable to drinking from a water fountain or glass. Sales are increasing in the United States at a rate of 15 percent each year, and larger companies are monopolizing this profitable industry.

ESSENTIAL USAGE

In addition to the luxury usage, bottled water can become an essential commodity at times of outbreaks of waterborne bacterial and parasitic disease due to contaminated public water supplies. Also, when private wells or community water supplies become dangerously contaminated by pesticides, fertilizers or other chemicals in rural areas, the availability of bottled water is important. In emergency situations of groundwater and surface water contamination, such as floods and earthquakes, bottled water is in great demand and serves as a short-term solution to the problem until normal water supplies can be restored.

The reliance on bottled water at times of emergency suggests that the product is safer and less of a health risk than water from the faucet. It is not generally recognized that one-third of all bottled water in the United States comes from public water supplies and is no safer than the water obtained from the faucet. Furthermore, the regulations governing the purity of bottled water may be less stringent than those applied to ordinary tap water.

An Environmental Policy Institute Report (1989) refers to the frequency of contamination of bottled water with low levels of heavy metals, solvents, and bacteria. The report cites many independent studies concluding that neither the government nor the bottled water industry can offer consumers a guarantee that all bottled water is healthful and contaminant free. Despite the value of bottled water as an alternative source of drinking water at times of emergency, "consumers should not presume that bottled water is invariably preferable to tap water."

TYPES OF BOTTLED WATER

Bottled water products may be classified into two main caregories:

1) Nonsparkling or "Still" bottled water; and
2) Sparkling or "Carbonated" bottled water.

STILL BOTTLED WATER
The still or non-carbonated form of bottled water is used primarily for drinking purposes, in cooking,

and for ice cubes. More than 90 percent of bottled water consumed in the United States is nonsparkling or the still variety. It is available in 5-gallon containers for delivery to homes or offices, or in one gallon, one liter, or smaller bottles purchased in supermarkets and drug stores. The average cost of a liter bottle is 90 cents.

Within the "still" water category there are a number of types that identify the source or treatment of the water before bottling. These types are defined under proposed FDA regulations as follows:

• **Spring water** flows naturally to the surface from an underground aquifer, or it may be collected underground through a bore hole where a spring emerges;

• **Artesian water** is drawn from a well drilled into a confined aquifer (a water-bearing rock or rock formation) containing water under pressure and situated above the natural water table, so that a pump is not required;

• **Well water** is pumped from a well drilled, bored, or otherwise constructed in the ground to tap into a ground water source or aquifer;

• **Natural water** refers to any bottled spring, artesian, well, or mineral water derived from a ground water source and not from a municipal water supply system. Natural water may or may not need to be filtered or treated with ozone or other disinfection process.

Natural spring water contains minerals that vary in concentration according to the brand. Most are low in sodium or sodium free. The label should be checked for the type and content of minerals. Calcium, magnesium, and bicarbonate are in highest concentration, and nitrates are absent or minimal;

• **Mineral water** comes from a source tapped at one or more bore holes or springs originating from a geologically protected underground water source. As defined by the IBWA and as required by several states, mineral water should contain at least 500 ppm of total dissolved solids. Most but not all mineral waters are carbonated;

• **Distilled water** is still water that has been treated by distillation - vaporizing and then condensing water to free it from dissolved solids;

• **Purified water** is still water that has been treated by distillation, deionization, reverse osmosis or other process to meet the requirements of purified water as defined in the most recent edition of the U.S. Pharmacopeia. Purified or distilled water is used in laboratories and for medical purposes. It is sometimes recommended by physicians for patients on restricted diets, such as low sodium diet for high blood pressure;

• **Nursery drinking water** is a distilled water to which minerals, including calcium and magnesium chloride, potassium bicarbonate, and sodium fluoride, have been added;

• **Drinking water** comes from any approved source, including the public water supply, provided it and has undergone disinfection and filtration processes and meets the contaminant standards of the Safe Drinking Water Act.

• **Flavored bottled waters** of any type have to comply with the same maximum allowable contaminant levels required of other bottled waters.

CARBONATED BOTTLED WATER

Carbonated or sparkling water can be obtained directly or "naturally" from a spring or well or it can be produced by carbonation of still water. Pressurized carbon dioxide is forced through the water to form carbonated water.

Carbonated water is usually a mineral water and is characterized as effervescent or bubbly. It contains not less than 500 ppm of total dissolved solids (TDS). Some brands contain more than 2,500 ppm of TDS. The sodium content can vary between 40 and 400 ppm. For those on a low sodium diet for health reasons, the label should be studied carefully.

Health conscious consumers are impressed with the named brands, such as "Naturally Sparkling Mineral Water," or "Sparkling Natural Mineral Water." Carbonated bottled water is enjoyed as a refreshment beverage, as an alternative to soft drinks, coffee and alcoholic beverages In contrast to soft drinks and alcohol, sparkling waters have no calories, artificial sweeteners or caffeine. Some contain an essence of flavor. The cost of a liter bottle of an imported mineral water can be as high as $1.79.

Bottled mineral water imported from Europe is regulated by the country of origin. In the United States, mineral water was exempt from regulation by the FDA until 1993. Now, mineral water as well as flavored bottled waters have to comply with the same maximum allowable contaminant levels required of other bottled waters. These regulations, however, exclude products labeled as "carbonated" water, "seltzer" water, "soda" water, and "tonic" water, because they are considered soft drinks. Some states require that these waters meet

FDA quality standards.

Soda water, club soda, and seltzer water are different names for the same bottled water product. Soda water is made by introducing carbon dioxide under pressure into potable water. "Potable" means that the water is suitable for drinking and meets the Safe Drinking Water Act regulations. Prior to the era of bottled soda water, some households would make their own soda water using a carbon dioxide cartridge and a soda water siphon bottle.

QUALITY CONTROLS

Unlike public water supplies which are regulated by the U.S. Environmental Protection Agency (EPA), bottled water is regulated by the Food and Drug Administration (FDA) and is considered a "food." The manufacturers and marketers of bottled water must comply with the Federal Food, Drug and Cosmetic Act (FD&C Act), and when sold as a consumer commodity, bottled water is subject to the Fair Packaging and Labeling Act (FPLA). The FDA issues quality standards to assure the safety of bottled water.

BOTTLED WATER STANDARDS

Bottled water quality standards were originally adopted in 1973 and were based on the 1962 U.S. Public Health Standards for drinking water. The Environmental Protection Agency (EPA) was made responsible for the safety of municipal water systems in 1974, under the provisions of the Federal Safe Drinking Water Act (SDWA).

The SDWA set maximum limits for bacteria,

chemicals, and radioactivity that are health hazards, and for physical contaminants that affect the aesthetic qualities of drinking water, including odor, taste, and color. Amendments to the SDWA in 1986 and since then required the EPA to set additional standards.

FDA Role. The FDA has primary responsibilty for the safety and quality of bottled water. The FDA is required to follow the EPA additional or amended standards for contaminants in municipal drinking water by setting acceptable standards for bottled water. If the FDA chooses not to follow the EPA standards, it must publish its reasons for not doing so in the *Federal Register.*

Since 1975, the FDA has been responsible for ensuring the quality standards for bottled water, in conformity with EPA standards for tap water. Maximum contaminant levels for pesticides, mercury and radionuclides were included in the FDA standards in 1978.

Standards proposed by the FDA for seven volatile synthetic organic chemicals, including benzene, carbon tetrachloride, trichloroethylene, and vinyl chloride, were published in the Federal Register on July 6, 1990. These proposed FDA standards shown in Table 10-1 followed new EPA regulations for maximum contaminant levels (MCL) of the seven chemicals. They were finally adopted by the FDA January 1993. The bottled water standards do not apply to sparkling mineral water or soft drinks, including soda water.

TABLE 10-1. FDA Standards for seven Volatile Synthetic Organic Chemicals proposed July 6, 1990 and adopted Jan 3, 1993.

Chemical	*MCL (ppb)*
Benzene	5
Carbon tetrachloride	5
1-2-dichloroethane	5
1,1-dichloroethylene	5
1,1,1-trichloroethane	200
Trichloroethylene	5
Vinyl chloride	2

Reproduced from FDA Talk Paper, August 23, 1990.

Most of the water bottled at the 475 plants in the United States should meet the proposed new quality standards without further processing. About 70 percent of bottled water comes from private water sources that are protected, and they are unlikely to be contaminated with synthetic organic chemicals from industrial wastes. In a 1986 survey of bottled water in New York, all 93 products tested complied with the MCLs of volatile organic chemicals (VOCs).

Bottled water that is derived from municipal sources should be in compliance with EPA standards for VOCs. Some states have their own standards and permitted maximum levels for organic chemicals in bottled water.

FDA requires that bottled water be safe and clean, and that sanitary precautions be used in the processing and distribution of bottled products. The FDA is responsible for inspecting bottled water facilities regularly. All bottled waters, including mineral water, must be produced according to FDA Good Manufacturing Practices. The FPLA should guarantee honest and informative labeling.

Under the Good Manufacturing Practices regulations, FDA requires sampling and analysis of the source water and each bottled water product for chemical contaminants at least once each year. The regulations also require the sampling and analysis, after processing but before bottling, to assure uniformity and adequacy of the processing performed by the manufacturer.

In January 1993, the FDA proposed standard definitions for all bottled water products, to avoid confusion in labeling and differences among states. These definitions are outlined under the section on types of bottled water. New limits were established for 50 chemical and other contaminants, in addition to 31 previously regulated.

The MCLs for lead, copper, mercury, barium, and cadmium were revised, and levels for 28 synthetic organic chemicals, pesticides, and polychlorinated biphenyls were established or modified in these 1993 FDA proposals.

EPA Role. The EPA has primary responsibility for the safety of drinking water supplied by public water systems. The EPA determines standards of quality for municipal drinking water and expects the FDA to adopt the same or alternative acceptable standards for bottled water.

State Role. States may adopt their own regulations which are sometimes stricter than FDA requirements. Many states (e.g. California, New York, Pennsylvania, Massachusetts, Florida) had already included testing for the seven VOCs in bottled water that were added to the FDA regulations in 1993. State agencies, some under contract to the FDA, and industry conduct regular inspections of bottled water plants.

Industry Role. The International Bottled Water Association (IBWA) represents U.S. and international companies that produce bottled water. It plays an active role in bridging the gap between the industry and the FDA. In 1988, the IBWA petitioned the FDA to establish stricter quidelines for bottled water. Their program of self-policing the industry is important in maintaining the quality and safety of bottled water.

Unfortunately, FDA's regulations do not require these self-monitoring results to be reported. These apparent shortcomings of cross-checks of compliance with FDA regulations have led to criticism of the FDA role by the Environmental Policy Institute, a public interest organization based in Washington, D.C.

BOTTLED WATER CONTAMINANT RISKS

Approximately 75 percent of bottled water comes from springs and wells that are protected, and are at less risk of contamination by microorganisms, synthetic organic chemicals, and pesticides than surface water systems. The other 25 percent derived from municipal water systems must meet federal and state regulations according to the Safe Drinking Water

Act.

In terms of source, bottled water should at least be as safe as municipal water, and that coming from protected sources should be even safer. Why then did the Environmental Policy Institute (EPI) conclude in a 1989 report that bottled water, in general, is not necessarily any safer or more healthful than the water which comes from most faucets?

Factors which led the EPI to question the presumed safety of bottled water included the following:

 • EPA has not kept pace with the need for more drinking water standards;

 • FDA has failed to adopt the newly established EPA standards for eight volatile organic contaminants that may be found in bottled water, and has failed to require bottled water suppliers to test for up to 51 additional chemicals which EPA monitors at public water sytems;

 • FDA's requirement for bottled water to come from approved sources is not forcibly regulated;

 • The consumer is not sufficiently informed of the fact that 25 percent of bottled water comes from public water systems;

 • Various studies have shown that bacterial and chemical contaminants are periodically detected in bottled water at levels in excess of health standards, and low levels of various chemicals are reported more frequently. The EPI concludes that bottled water is not entirely "pure" or "contaminant-free," as suggested by industry advertising;

 • The self-monitoring program conducted by the manufacturers, as approved by the FDA, does not include reporting requirements or unannounced inspection of records;

• Mineral waters were exempt from FDA quality standards and have been found to contain potentially harmful levels of chemicals and other contaminants;

• Some contaminants found in bottled waters are introduced during processing and storage in plastic containers. A carcinogen, methylene chloride, may enter bottled water from the polycarbonate resin in certain plastic bottles, and bacteria may multiply during prolonged storage, according to the EPI report.

Many of the criticisms of the FDA regulations which could have applied in 1989 have since been corrected. As stated earlier in this chapter, under a final rule published in the Jan. 3, 1993 Federal Register, the FDA complied with the EPA standards for levels of seven synthetic volatile organic chemicals in bottled water. It also established or modified permitted levels for other chemical contaminants, and applied the quality standard requirements for bottled water to include mineral water. The FDA regulations for bottled water are now more closely equated to those used by the EPA for municipal water supplies.

Bacterial Contaminant Studies. Studies cited by the Environmental Policy Institute included the following:

• EPA scientists led by Dr Edwin Geldreich found in 1975 that the level of microbial contamination of bottled water increased significantly during storage, and refrigeration retarded growth of the bacteria. Many bottled waters met the standard for coliform bacteria but contained high levels of opportunistic organisms, capable of causing disease in infants and adults with suppressed immune systems.

• Researchers from the University of Cincinnati studied the effect of storage on the microbiological quality of bottled water (Scarpino et al, 1987). More than one quarter of the samples of domestic bottled water tested had high bacterial counts after 48 hours storage at 95 degrees F. Storage in a warm environment may facilitate bacterial contamination and pose a health risk.

• A study of the bacteriological quality of 58 different brands of still bottled water and carbonated European mineral water at the University of Wales, U.K., found 19 percent tested positive for staphylococcus, an organism probably introduced by handlers during the bottling process. (Hunter and Burge, 1987).

Chemical Contaminants. Studies performed in the States of California and New York, and cited by the EPI, have found chemical contaminants and irregularities in bottling plants that provoked the enactment of more stringent standards for bottled water in some states:

• A 1985 report from the California State Assembly's Office of Research, based on consumer complaints and inspection results, found citations of bottled water plants for adding chlorine to test samples to kill bacteria, for failure to monitor their water sources or keep records of water testing results, and for falsifying records. Concentrations of chemicals above the EPA and FDA standards, reports of foreign materials, and water sources located in proximity to chemical contamination sites were some of the complaints filed.

• A 1982 report on bottled water prepared by the New York, Suffolk County Department of Health Services, found that one-third of the distilled waters examined contained chloroform and other organic chemicals, most frequently trihalomethanes. Almost 80 percent of mineral, seltzer and soda waters showed trace amounts of organic contaminants, and some had excess sodium levels. The New York Department of Health recommended that bottlers should identify the source of these organic contaminants and eliminate them. They should also be required to disclose on the label any excess sodium levels.

• In contrast to the adverse reports concerning bottled water in California and New York States, the State of New Jersey survey of 41 brands produced and/or sold in the state found that 38 of the samples met the New Jersey standards. They were of high quality and exceeded both the bottled water and community water standards for levels of volatile organics and total coliform organisms.

State of Wisconsin sampling and analysis of test results in 1987 showed no evidence of excess selected trace metals or organics, and coliform levels were within acceptable health standards. However, the testing methods used in the Wisconsin study were thought to be less sensitive than those employed in California and New York.

These reports, emphasized by the EPI as cause for concern regarding the safety and quality of bottled water products in the United States, are indications for closer FDA and state inspections of water sources and bottling procedures. Since FDA regulations were modified in 1993 and more of the EPA standards for public water supplies have been applied to bottled water, the consumer can be more confident in the advertised purity of the various products. Future reforms to be enacted in Congress may weaken this confidence, however.

Projected Congressional Reforms. Reform of the Safe Drinking Water Act is expected to be one of the first environmental health priorities in the new Congress in 1995. Rather than an expansion of EPA and FDA regulations of drinking water, cuts and relaxations are likely.

The Environmental and Energy Study Institute, in its Weekly Bulletin, Jan. 2, 1995, predicts the following issues to be considered by Congress:

• The EPA may be asked to reduce costs by relaxing the current standard setting policy, especially for small water supply systems, which calls for the use of best available technology to achieve maximum health protection.

• Given the pressure to cut spending, the proposed new funds for states and localities for testing and inspections of water systems may be cut or eliminated.

• Provisions to force small systems to consolidate may be introduced, in order to reduce costs, and certification requirements for treatment facility operators may be dropped.

Cost-cutting proposals and relaxations of the

types outlined will receive strenuous opposition from environmentalist groups. Compromises and cost-benefit analyses may provide relief from expensive testing and monitoring requirements for small systems, while maintaining essential health safeguards for all drinking water supplies.

Plastic Bottle Contaminants. The chemical migrants from plastics of particular concern to health authorities are vinyl chloride poymers (PVC), acrylonitrile polymers (ABS), and nitrosamines. All have been found to cause cancer in animals and some in humans.

Increased levels of methylene chloride, another carcinogen, were found in a California study cited by the EPI. In 1986, the California Department of Health Services studied the effects of storage of bottled water in plastic containers. After 14 days, the concentration of methylene chloride in the water was 4 to 20 ppb and far above the standard limit of 1 ppb required in California.

It may surprise consumers to realize the enormous potential for risk of intoxication from a multitude of migrant chemicals contained in plastic containers. It is equally reassuring to learn of steps taken by the FDA, the EPA, and other governmental authorities to protect the consumer from these hidden hazards. For example, enforced improvements in technology effected dramatic reductions in vinyl chloride in foods, and the use of acrylonitrile copolymer bottles for packaging soft drinks and other beverages was banned by the FDA in 1977.

In accordance with the World Health

Organization Committee recommendations on Food Additives, the dietary exposures to vinyl chloride, acrylonitrile, and other packaging materials is now controlled in several countries. From a desire by trading countries to eliminate international barriers, several initiatives have been taken to control food and water packaging materials and their potential health risks.

With the knowledge that some undetected toxic chemicals may concentrate in bottled water after prolonged storage, it is important to check the bottling and expiration dates on products before purchase. The following suggestions for consumers may provide tips for buying bottled water - When to buy, What to buy, and What precautions to take:

SUGGESTIONS FOR CONSUMERS

WHEN TO BUY

• If your drinking water suddenly becomes turbid, changes in color or taste, or an unpleasant odor develops.

• If you learn that the public water system has become contaminated and you are instructed to boil the water. Bottled water is a simpler alternative.

• At times of floods or earthquakes when water contamination is a common hazard.

• If your home has lead pipes or lead solder, drink bottled water until your water has been tested. Also, cook with bottled water, since lead may concentrate during extended preparation of soups and stews.

• If you have a private well for drinking water

and you suspect contamination with coliform bacteria or nitrates. Drink bottled water until the well water is tested.

• If your neighborhood has a cluster of cancer patients and water contamination from a landfill is suspected, bottled water may be preferable to community water supplies or private wells.

• If you need to limit your intake of sodium for medical reasons, bottled waters of low sodium content or distilled water may be preferred to public supplies.

• If you are travelling and suspect contamination of local water supplies, drink only brand named bottled waters in sealed containers.

• If your water tastes strongly of chlorine, bottled water is preferable for cooking as well as drinking. Bottled water does not contain chlorine and does not taint the food.

WHAT TO BUY

• Buy water from approved protected sources, bottled at the source, and well known and respected brands.

• If the bottler uses water from a public source, try to determine the locality of the original source or sources and its record of safety. Spring or ground water is usually safer than surface or stream water. Consult an EPA or a state water official.

• Is the bottler a member of the International Bottled Water Association?

• Look at bottling dates and expiration dates. Buy the product most recently bottled.

• Read the labels for mineral content, especially sodium. If you are on a low sodium diet, consult your doctor about distilled water.

• Don't feed small infants distilled water without your doctor's approval. Choose nursery water, a distilled water with added minerals.

• Most bottled waters are lacking in fluoride. Check with your dentist if your child is drinking only bottled water.

• Remember that seltzer and club soda are soft drinks and are not subject to FDA bottled water quality standards. Some soft drinks contain sugar and sodium, whereas bottled water does not.

WHAT PRECAUTIONS TO TAKE

• Inspect the bottle for a seal and any obvious foreign material.

• Don't drink the water if it has an unpleasant odor or taste. Contamination with cleaning fluids has occurred during the bottling process, even with well known brands. You should not be able to detect a chlorine taste in bottled water.

• Don't buy more than you can drink in a week. If you have to store the water, place it in a refrigerator. Bacteria grow more rapidly in warm than in cold water. Refrigerated water tastes better than water at room temperature.

• Do not assume that bottled water is always safer than tap water.

• Call the International Bottled Water Association (800-WATER11) if you need further information about bottled water quality and safety.

Environmental and Energy Study Institute. Safe drinking water a high priority. Weekly Bulletin, Jan 2, 1995:9.

Geldreich E et al. The necessity of controlling bacterial population in potable waters - Bottled water and emergency water supplies. J Amer Water Works Assoc. March 1975:117.

Hunter PR, Burge SH. The bacteriological quality of bottled natural mineral waters. Epidemiol Infect. Oct 1987;99:439-443.

Lambert V. Bottled water: New trends, new rules. FDA Consumer June 1993:9.

Markussen K, Bonczek R. Organic chemical compounds in bottled water products distributed in New York State. New York State Department of Health, Bureau of Public Water Supply Protection, 1982.

Marquardt S et al. Bottled Water: Sparkling Hype at a Premium Price. Washington, D.C., Environmental Policy Institute, 1989.

Scarpino PV, Kellner GR, Cook HC. The bacterial quality of bottled water. J Environ Sci Health May 1987;A-22:357-367.

State of California. Bottled water and vended water: Are consumers getting their money's worth? California State Assembly, Office of Research, March, 1985.

State of New Jersey. New Jersey bottled water survey. New Jersey Department of Health, Environmental Health Services Unit, 1986. Quoted in Environmental Policy Institute publication on Bottled Water by Marquardt S et al, 1989.

State of New York. 1982 Report on bottled water and bottled water substitutes. Suffolk County Department of Health Services, Drinking Water Supply Section, April 1982.

State of Wisconsin. Bottled drinking water sampling and analysis test results for fiscal year ending June 3, 1987. Madison, Wisc, Wisconsin Department of

Agriculture, Division of Trade and Consumer Protection, 1987.

Warburton D et al. Microbiological quality of bottled water sold in Canada. Can J Microbiol. 1986;32:891.

CHAPTER 11

COLOR, ODOR, TASTE, AND HARDNESS

C olor, odor, taste, and hardness are the aesthetic qualities of drinking water that do not necessarily present a health risk to the consumer. However, water that appears dirty or cloudy, and has an unpleasant odor and taste will undermine consumer confidence and lead to the use of bottled water as an alternative to the public water supply.

The EPA has developed a list of Secondary Drinking Water Standards containing 13 contaminants that affect the aesthetic aspects of water, as shown in Table 11-1. These suggested levels of contaminants are recommended to the States as guidelines and reasonable goals, intended to protect public welfare, but federal law does not require water systems to comply with them. States sometimes adopt their own enforceable regulations governing these qualities of drinking water.

TABLE 11-1. National Secondary Drinking Water Standards, EPA 1991.

Contaminants	Levels	Effects
Aluminum	.05 - .2 mg/L	discoloration
Chloride	250 mg/L	taste, corrosion
Color	15 units	aesthetic
Copper	1 mg/L	taste, staining
Corrosivity	non-corrosive	leaching lead
Fluoride	2 mg/L	brown teeth
Foaming agents	.5 mg/L	aesthetic
Iron	.3 mg/L	taste, staining
Manganese	.05 mg/L	taste, staining
Odor	3 threshold	aesthetic
pH	6.5 - 8.5	corrosion
Silver	.1 mg/L	discolored skin
Sulfate	250 mg/L	taste, laxative
Total solids (TDS)	500 mg/L	taste, corrosion
Zinc	5 mg/L	taste

Adapted from US EPA Publication, September 1991.

Some of these contaminants discolor the water, teeth, skin, porcelain, or laundry (aluminum, fluoride, silver, copper, iron, and manganese), some cause an unpleasant taste (chloride, copper, iron, manganese, sulfate, and zinc), and the total dissolved solids (TDS) affect hardness and the corrosivity of water.

COLOR OF DRINKING WATER

Color in drinking water is undesirable and is due mainly to dissolved organic substances derived from decaying vegetation and humus. Other substances that may contribute a color to ground and surface waters include iron and manganese, copper, and industrial wastes from pulp, paper and textile industries. The iron and copper may also be derived from the plumbing and water pipes.

Rarely, so-called "iron bacteria" cause a red coloration due to oxidation of iron to form an insoluble ferric hydroxide salt, and dissolved manganese can be precipitated as an insoluble oxide by the action of bacteria, causing a black discoloration. These color problems of bacterial origin affect ground water more frequently than surface waters. They may be severe enough to block distribution lines.

Color is removed from natural water by coagulation, filtration, and chemical oxidation. Colors above the EPA suggested limit of 15 color units are usually detected by the consumer. Removal of color prior to chlorination limits the production of trihalomethane by-products and the undesirable substances attached to humus. The WHO recommended guideline value for color in drinking water is less than 15 true color units (TCU).

TURBIDITY

Turbid or cloudy water is due to finely divided solid particles that are suspended in the water and do not dissolve. Surface water from lakes and streams often contains soil particles derived from eroded land and run-off. Ground water and deep wells are rarely affected by turbidity.

The particles consist of clay, organic matter from decomposition of plants, and fibrous materials such as asbestos. Microorganisms that can cause turbidity include the blue-green algae that bloom in the summer, and iron bacteria that form red-water in distribution systems. Turbidity protects bacteria and viruses from treatment with disinfectants, and turbidity must be removed before chlorination to obtain water safe to drink.

The major city water supply contaminated by *Cryptosporidium* in a recent outbreak affecting nearly a half a million consumers was noted to be turbid before the cause of the epidemic was determined. An educated public would have discarded the turbid water as a potential health risk and substituted bottled water. Unfortunately, they drank the water, not realizing the dangers attending the appearance of turbidity.

Removal of turbidity is accomplished by filtration, or by a combination of coagulation, sedimentation, and filtration. The efficiency of coagulation is temperature dependent and is increased in warmer water. The true color of water can only be assessed after turbidity has been removed.

The ingestion of turbid water that has been chlorinated may be a dangerous health risk. Turbidity should be kept below 1 NTU, if possible. Nephelometric turbidimeters measure the intensity of light in NT units scattered at 90^0 to the path of a beam of light directed through the sample of water. A level above the WHO guideline value of 5 NTU is generally detected by consumers and is not only objectionable but also, a health risk.

ODOR

Objectionable odors in drinking water may be of natural or industrial origin. Odors of natural or biological origin arise from pathogenic organisms and from dumping of raw sewage into the surface water sources. They are earthy, musty, or sour, and sometimes, fishy, or grassy. The musty odors may be caused by the actinomycetes group of organisms.

Industrial odors are associated with pollution from commercial waste, and often smell like petroleum, creosote, naphthalene, or phenol. Some chemical contaminants, like cyanide, can be detected by the sense of smell more acutely than by analytical methods, whereas the odors of pesticides are too weak for detection at their recommended guideline values.

Drinking water should be free of odor as judged by at least 90 percent of the population. In some communities, a perceptible chlorine odor is regarded as acceptable and a sign of a risk-free water supply.

TASTE

Taste and odor problems account for the majority of consumer complaints about drinking water. Surface waters are affected more frequently than ground water, and seasonal variations in taste and odor due to biological factors are not uncommon. Temperature is an additional variable; the growth of microorganisms which may affect taste is enhanced in summer months when raw water is warmer, and taste intensity is greatest at room temperature.

The taste and odor of chlorine in water is objectionable to some consumers, but provides a sense of security to others, particularly when living in developing countries where bacterial contamination of drinking water is common. A number of factors can influence the taste threshold for chlorine, including the pH (alkalinity increases the taste threshold), and the temperature (increased temperature raises the level of free chlorine and lowers the taste and smell thresholds). The pH also influences the equilibrium and threshold for sulfide, or rotten egg odor and taste of some contaminated water supplies.

Many inorganic constituents in water cause an unpleasant salty taste. The taste thresholds for magnesium, calcium, sodium, and potassium range from concentrations of 30 to 300 mg/L in water. The taste thresholds for iron and zinc in distilled water are lower than those for mineralized spring water. The total dissolved solids in spring water have a protective effect, and some unpleasant tastes are disguised.

The taste of water is assessed by panels of laboratory technicians, using a five point scale (good - not observed - weak - objectionable - bad), or by a "forced choice method," using a taste number, acceptable below 1. A choice must be made between paired samples, including taste and odor-free controls, even when no difference is detectable. A wide variation in individual taste and odor detection is observed among consumers. Threshold concentrations are those detected by 50 percent of a panel of judges.

THE TEMPERATURE FACTOR

Water with ice tastes better than water straight from the faucet. The intensity of taste is reduced by chilling water for drinking. The rates of chemical reactions in water will decrease with decreasing temperatures. The physical, microbiological, and chemical aspects of water are all affected by temperature.

Physical Changes. Turbidity and color are affected by the coagulation process which is temperature-dependent. As temperature decreases, the rates of sedimentation and filtration are decreased. The efficiency of turbidity removal is impaired under winter conditions. In contrast, activated carbon filters work best with cold water.

Bacterial Contamination. Temperature has effects on water treatment processes, especially disinfection. The efficiency of chlorination and ozone treatments are aided by an increase in temperature of the water. Both bacteria and viruses are killed more readily by disinfection when the temperature is raised from 5^0C to 25^0C.

Coliform bacteria survive longer in raw water in winter months, but the growth of "nuisance organisms," such as algae that block strainers and filters, is increased in summer. Viruses survive longer than bacteria, but waterborne hepatitis A infection has not been correlated with raw water temperature. The survival of parasitic cysts and ova is favored by cold.

Chemical Changes. The formation of chlorine by-products, such as trihalomethanes, is influenced by the season of the year, and is increased in the warmer months. Corrosion is also a greater problem during the summer. The solubility of calcium carbonate and the total dissolved solids (TDS) are also affected by the temperature of water.

TOTAL DISSOLVED SOLIDS

In addition to carbonate ions, the TDS consist of bicarbonate, chloride, sulfate, nitrate, sodium, potassium, calcium, and magnesium. The TDS contribute to the taste, hardness, and corrosion and incrustation qualities of water. They originate from natural sources, sewage, urban runoff, and industrial waste discharges. The use of salt for snow and ice control in winter contaminates both surface and ground water supplies, leading to high TDS values.

The palatibility of water is improved by concentrations of TDS up to 500 mg/L (the maximum level suggested by the EPA), but above this concentration, the taste of water becomes less palatable and unacceptable. Although no adverse health effects have been correlated with excess levels of TDS, a WHO health-based guideline value of 1000 mg/L, proposed in 1984 and dropped in 1993, should not be exceeded. Total dissolved solids are not removed by conventional methods of water treatment, and their effects on corrosion and incrustation in water distribution systems may lead to increased lead levels and damaged plumbing.

HARDNESS

Hardness of water is due primarily to calcium

and magnesium. A unit of hardness is expressed as mg of calcium carbonate equivalent per liter. Water sources rarely contain more than 500 mg/L, and drinking water hardness ranges from 10 to 500 mg/L. Their are two types of hardness, carbonate (temporary) and non-carbonate (permanent). Hardness is a measure of the ability of water to react with soap. High hardness causes scum formation and scale deposits. Low hardness, when associated with a paucity of total dissolved solids, may lead to increased corrosion. The lack of all minerals, not just calcium or magnesium, causes water to be corrosive.

Health Effects. Apart from studies that suggest a link between excessive hardness of water and kidney stones, there is no clear evidence of ill health caused by water hardness. Studies correlating hardness with congenital nervous sytem defects, infant mortality rates, and some cancers are of doubtful significance. Despite earlier reports that water hardness may protect from heart disease, more recent studies have failed to correlate water hardness with cardiovascular disease mortality rates.

Hammer DI, Heyden S. Water hardness and cardiovascular mortality. JAMA June 20, 1980;243:2399-2400.

United States Environmental Protection Agency. Is Your Drinking Water Safe? Washington, D.C., EPA Office of Water, September 1991.

World Health Organization. Guidelines for Drinking-Water Quality. Vol 2. Health Criteria and Other Supporting Information. Geneva, WHO, 1984.

Idem. Guidelines for Drinking-Water Quality. 2nd Edition. Vol 1: Recommendations. Geneva, WHO, 1993.

CHAPTER **12**

CONSUMER CONCERNS AND HOME WATER TREATMENTS

Taste and odor problems and concerns about the level of lead in drinking water are the most frequent questions asked by consumers. Is our water safe to drink? Do we need to substitute bottled water? Should we install a home water treatment unit, and if so, which type of unit would be most satisfactory for our particular needs? This chapter attempts to cover the more common aesthetic quality complaints and health related contaminants that may be corrected or removed by home treatment units.

The Environmental Protection Agency (EPA) provides information in response to inquiries about the necessity for home treatment units. It does not regulate the manufacture, distribution, or use of these units. The EPA stresses that home treatment of water for health protection and removal of hazardous contaminants is rarely necessary. The purpose of home treatment units is to improve the aesthetic and cosmetic qualities of water, especially taste, color, and odor.

Public Water Systems. The 85 percent of Americans who receive their water from public water systems are usually assured of a safe supply. If the local system has failed for some reason, the authorities are expected to notify the consumer and advise about protective measures, such as boiling the water or substituting bottled water, until the health-based standards are met. Unfortunately, the notification of the public in some emergency situations may be delayed, and an outbreak of disease of major proportions may develop.

Consumers should be aware of changes in the quality of their water supply (eg. turbidity, change in color, or unpleasant odor) that might herald contamination with certain organisms or chemicals with health-risk potential. Their immediate response and substitution of bottled water could protect them from possible serious illness. For example, the recent epidemic of waterborne *Cryptosporidium* infection in the city of Milwaukee was heralded by an acute change in appearance of the water from the tap manifested as turbidity.

Private Well Water. The 15 percent of Americans who obtain their water from private household wells are primarily responsible for the quality of their own water supplies. They are subject only to state and local laws, but not federal regulations. Local health departments will test wells for bacterial contamination on request, and sometimes nitrates, and will advise on the proper placement of a well in relation to the septic system. (See Chapter 9).

The owner must be the judge of the safety of the well water and any treatment needed to adequately protect himself and his family. He should identify potential sources of contaminants, including pesticides and fertilizers used in lawn care and agriculture, improper disposal of household chemicals and motor oil, and leaky, underground oil storage tanks. Advice on possible pollutants of wells and their management may be obtained on the EPA Safe Drinking Water Hotline (1-800-426-4791), from State Water Supply Agencies, or from the local health department.

COMMON CONSUMER CONCERNS

In addition to turbidity, taste, and odor, the color of drinking water, laundry stains, soap residue, bathtub ring, scale build up in plumbing, water heaters and humidifiers, and corrosion are frequent causes for consumer concern. A national poll of America's restaurants, sponsored by the Water Quality Association (WQA) in 1992, revealed that 50 percent of the chefs, owners, and managers surveyed were concerned about the quality of drinking water. Eighty-six percent of restaurants received their water from municipal systems and 10 percent from private wells.

Restaurant Owner's Complaints. The problems encountered in America's restaurant water supplies, in decreasing order of prevalence, were hardness (49%), odors (34%), chlorine taste and odor (33%), discolored water (16%), and poor taste (15%), according to the WQA report.

These deficiencies in the aesthetic quality of the water caused maintenance problems with kitchen appliances; spotting of china, glassware, and flatware; and complaints about the taste of coffee, tea, and ice cubes. The chief reason for installing a water treatment unit was the spotted china and glassware, not consumer complaints. The WQA found that water-related problems in restaurants were common in any area of the United States, and involved coffee shops and expensive dining establishments alike.

When a problem is suspected in the home, restaurant, or office water supplies, the water should be tested to determine the nature of the offending constituent.

Water Analysis. For aesthetic problems like hardness, the EPA recommends analysis by a water specialist certified by the Water Quality Association, Lisle, Illinois (1-708-505-0160). For contaminants with health risks, a laboratory certified by the State should be used for drinking water testing. The contaminant of concern (eg. lead) should be specified, when possible, because a full, quantitative analysis of inorganic and organic chemicals in water is expensive.

Having determined the nature of the problem or contaminant in the water, the choice of home treatment unit is important. No one treatment unit is a panacea for all possible problems. Two reliable sources on treatment units, endorsed by the EPA, are the Water Quality Association and the National Sanitation Foundation (NSF). The NSF develops and adopts standards and programs for consumer products, including home treatment units. The WQA provides advice on treatment units for specific uses in residential, commercial, industrial, and institutional establishments.

The EPA does not approve or endorse home treatment units, but certain filters must be registered by the EPA, if they claim to inhibit growth of bacteria.

EMERGENCY SAFETY MEASURES

If contamination of the drinking water is suspected and bottled water is not available, the following procedures may be used to disinfect the water to make it potable.

Boiling. Boil water for 15 minutes and allow to cool before drinking. Do not add ice since it may also be contaminated. Add a pinch of salt to each quart or liter, or pour the water from one clean container to another several times to improve the taste.

Chlorination. Add 10 drops (0.5 ml) of liquid chlorine household bleach to 1 quart or liter of contaminated water, stir or shake thoroughly, and allow to stand for 30 minutes. A chlorine odor should be detectable in the water. After disinfection period is completed, the chlorine content of the water can be reduced by heating or shaking to aerate the water, or allow it to stand for a longer time period.

Liquid chlorine laundry bleach usually has 4 to 6 percent available chlorine. Two to four drops per quart may be sufficient to provide a detectable chlorine odor. Read the label to make sure the bleach has chlorine and not hydrogen peroxide and check the percentage of chlorine available. If the available chlorine is only 1 percent and the water is cloudy, 20 drops per quart or liter will be required to provide a chlorine odor.

HOME TREATMENT UNITS

Controversies concerning the sale and use of home treatment units have arisen because of unsubstantiated claims of benefits and scare tactics. Consumers should beware of "free" water testing by a salesperson to determine the drinking water quality. An independent analysis is usually more accurate and meaningful. Check the manufacturers credentials, warranty, and maintenance provisions, and consult the WQA about your specific water problems and the appropriate type of treatment unit required.

Table 12-1 lists the types of home water treatment units available, and their specific abilities to address the various water quality concerns.

WATER SOFTENERS.

Hard water used in the home interferes with bathing, dishwashing and laundering. Soaps and detergents form a sticky soap curd on the skin, hair, dishes, glassware, and bathtub, and the life and lustre of linens and clothes washed in hard water are shortened and impaired. Scale formation in water heaters, boilers, pipes, and humidifiers reduces efficiency and leads to premature failure in performance.

The installation of a water softener in the cold water line where it enters the home can eliminate these problems. A process called *"ion exchange"* is used to remove the calcium and magnesium minerals which cause hardness. The hard water is passed through a column filled with a sodium containing resin, sodium permutit. The calcium element in the compounds causing the hardness is exchanged for the sodium ion in the resin. The sodium sulfate (Na_2SO_4) and sodium hydrogen carbonate formed by this ion exchange do not react like the calcium salts to form scale and scum. The calcium combines with the resin to form calcium permutit which is insoluble in water. The reaction is as follows:-

$$Ca\ SO_4 + Na\ Permutit \rightarrow Ca\ permutit + Na_2SO_4$$

TABLE 12-1. Home Water Treatment Units and Water Quality Concerns Benefited.

Treatment Unit	*Problems Benefited*
Cation Exchange	Water softener, lead, cadmium reduction
Anion Exchange	Nitrate reduction in wells
Activated Carbon Filters	Chlorine taste, odors, lead, mercury, asbestos, pesticide, volatile organic
Activated Alumina Filter	Fluoride, arsenic reduction
Reverse Osmosis	Taste, color, iron and manganese reduction
Oxidation	Iron, manganese, sulfide
Aeration	Volatile organics, radon
Mechanical Filtration	Turbidity, color
Microfiltration	Bacteria removal
Distillation	Dissolved solids removal
Disinfection[1]	Bacterial contaminants

[1]chlorination, ultraviolet, ozone, bromine, & iodine processes.

Adapted from Water Quality Association Answers, Sept 1994.

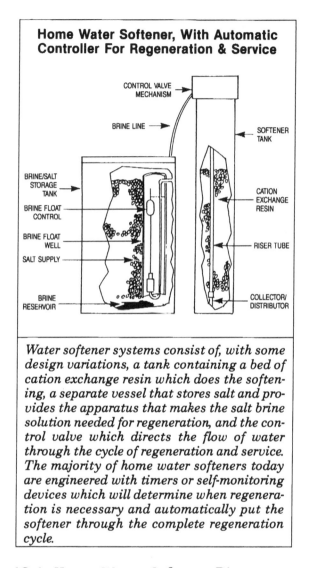

Home Water Softener, With Automatic Controller For Regeneration & Service

Water softener systems consist of, with some design variations, a tank containing a bed of cation exchange resin which does the softening, a separate vessel that stores salt and provides the apparatus that makes the salt brine solution needed for regeneration, and the control valve which directs the flow of water through the cycle of regeneration and service. The majority of home water softeners today are engineered with timers or self-monitoring devices which will determine when regeneration is necessary and automatically put the softener through the complete regeneration cycle.

Figure 12-1. Home Water Softener Diagram.
Reproduced with permission of the Water Quality Association from a Summary of "Research Report on Benefits of Soft vs. Hard Water in Laundering Operations" by Purdue University.

The hardness of water may be checked with a paper test strip obtainable from *Environmental Test Systems, Inc., P.O. Box 4659, Elkhart, Indiana 46514.*

Eventually, the supply of sodium on the resin is used up in this process, and the ion exchange material becomes exhausted. To reactivate or recharge the system, a strong solution of brine (common salt) is flushed through the bed of resin granules or beads. The salt drives out the accumulated hardness from the resin and replaces it with sodium. The excess salt is rinsed away with fresh water, and the the ion exchange column is then ready for further service.

This water softening-recharging cycle is repeated indefinitely and is operated automatically by a timer or a sensing device. The recharging process occurs every three days and lasts about two hours; it can be noisy and a disturbance to the whole household if it occurs during sleeping hours.

The homeowner adds salt to the system periodically to ensure a constant supply of soft water. The water softener capacity is given as the grains of hardness removed during a softening-recharging cycle. The capacity should be sufficient to last at least three days between cycles. Advice regarding the capacity required for a particular household should be available from the water treatment representative and dealer. The Water Quality Association validates tested equipment with a Gold Seal and provides information to consumers.

Diadvantages of Water Softeners.

Before investing in an expensive water softener, the homeowner should consider the pros and cons of removing calcium from the water. Hard water has certain advantages: 1) the taste of hard water is preferable to softened water; 2) the calcium in hard water is an essential mineral in the diet; 3) hard water forms a coating of insoluble sulfate or carbonate on plumbing and prevents the leaching of lead. However, these hard water properties are of doubtful importance (Water Quality Association, personal communication). From a health standpoint, levels of calcium in hard water are insignificant. In prevention of corrosion, the scale build-up in plumbing is porous, uneven, and ineffective. Water softeners do not cause or prevent corrosion, but naturally soft water is usually corrosive. Water softened by a treatment unit is not the same as naturally occurring soft water, in terms of corrosive properties.

Water Softeners and Septic Tanks.

Questions have been raised about possible adverse effects of water softeners on private sewage disposal systems. The volume of backwash and regeneration water discharged might overload the septic tank, causing solid wastes to be spilled into the drain field, and impairing the percolation of water through the soil. The salt-brine discharge could also kill bacteria important to the septic system.

A study conducted at the University of Wisconsin, Madison, and the National Sanitation Foundation, Ann Arbor, Michigan, and supported by the Water Quality Research Council, has concluded that water softeners pose no threat to the efficiency and safety of septic tank systems.

ACTIVATED CARBON FILTRATION

Activated carbon is an adsorbent material employed in household filtration systems to improve the quality of drinking water at the point-of-use.

Activated carbon filters reduce objectionable tastes and odors, such as chlorine, in water at the faucet. They also reduce levels of organic materials, including trihalomethanes, by-products of chlorination, some volatile organic contaminants, and certain pesticides and fungicides. The precoat solid block design of activated carbon provides 0.5 micron filtration as well as adsorption on very fine granules, capable of reducing lead and mercury in water.

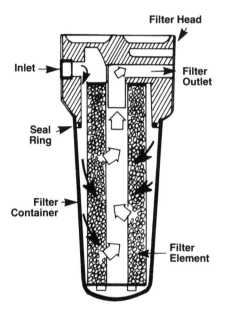

Figure 12-2. Activated Carbon Filter Diagram.
Reproduced with permission of the Water Quality Association.

The level of contaminant removal will vary with the size and type of filter, the degree of pollution, and the frequency of carbon renewal. Activated carbon filters will not remove all organic contaminants or all the lead in water, but they can be expected to reduce the levels of contaminants and lessen the health risks.

Types of Activated Carbon Filters.

These include the type attached to the faucet, pour through type, mounted under-the-sink, and whole-house filters. These filters are only to be used with municipally treated tap water or well water that is regularly tested for microorganisms and is safe to drink. They will not purify untreated raw water.

Some pour through filters have an ion exchange resin with silverized activated carbon, which inhibits growth of bacteria within the filter. They may be certified by the NSF international for lead and copper reduction, taste, odor, and chlorine reduction, particulate reduction, bacteriostatic effects, and zinc reduction. The filter has to be replaced every 35 gallons or 2 months, whichever comes first. The cost of replacement filters can be about $50 annually.

Whole-house filters are used to reduce water contaminants which may have adverse health effects during bathing or showering, or from inhalation. All activated carbon filters require periodic replacement of cartridges, and their use for all purpose water supplies may be expensive.

DISTILLATION

Point-of-use distillation will reduce unwanted minerals, chemicals, organisms, and tastes in water. Water is boiled, and the steam enters a condenser where it is cooled and returned as distilled water. The impurities and dissolved solids that are left behind in the boiler are drained away.

In addition to the reduction of unpleasant tastes

and odors, distillation reduces, lead, copper, sodium, nitrates, pesticides and fungicides, and volatile organic contaminants. Distilled water is different from water that has been boiled and cooled for drinking. Whereas distilled water is almost free of dissolved solids, such as sodium, nitrates, and scale-producing minerals, boiled water actually concentrates these minerals, including nitrates, and may magnify their health risks. Both distillation and boiling will kill microorganisms.

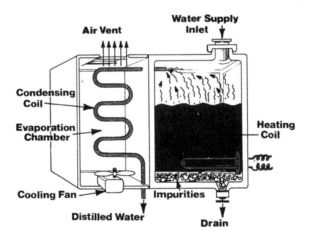

Figure 12-3. Household Distillation System Diagram. Reproduced with permission of the Water Quality Association.

Distilled water lacks many of the nutritionally important minerals found in hard water, such as calcium, magnesium, and sodium. It may be beneficial in patients with high blood pressure who require a low sodium diet, but in general, its continued use may impair bone and teeth development and calcification, with the risk of osteoporosis in adults. In young infants, the feeding of distilled water to replace fluids lost during bouts of diarrhea and vomiting has resulted in *water intoxication,* a serious neurological disorder with convulsions.

OXIDATION TREATMENT UNITS

Oxidizing filters are used to remove iron, manganese, and hydrogen sulfide, when these contaminants are too concentrated for a water softener to correct the problem. The Safe Drinking Water Act of 1974 suggested limits of 0.3 mg/L (parts per million) of iron and 0.05 mg/L of manganese. Brown to black stains build up in sinks and baths and on linens and other fabrics washed in water containing higher concentrations of these minerals.

The media in these filters will oxidize dissolved iron or manganese, converting it to an insoluble salt that can be removed by filtration. The deposits accumulated are removed from the unit by a periodic backwash and flush down a drain. The oxidizing action of the filter media is regenerated by the periodic addition of potassium permanganate to the unit.

If the iron or manganese is bound to organic matter, or if "iron and manganese bacteria" are present, a brown, slimy growth occurs in the toilet flush tank, and the oxidation method of removal is ineffective. A solution of hypochlorite bleach is introduced into the water system, which disinfects the water and allows oxidation of the iron and organic matter to occur. A carbon filter is then used to remove the precipitated particles and excess chlorine.

SEQUESTRATION

Another process for the treatment of water containing dissolved iron and manganese is sequestration, using a chemical feed pump. Polyphosphate compounds added to the water by the pump react with the iron and manganese and prevent their tendency to deposit. The minerals treated by this process will not stain or cause blocking of pipes and toilets.

REVERSE OSMOSIS

These point-of-use, under-the-sink treatment units force pressurized water through a semipermeable membrane which holds back contaminants suspended in the water and permits the passage of the purified water. The reverse osmosis membrane has microscopic pores which will not allow large molecules to pass. The size of the pore in the membrane determines the rate and efficiency of the unit.

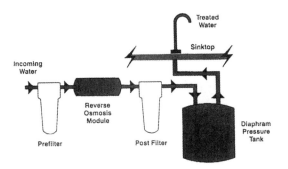

Figure 12-4. Reverse Osmosis Unit Diagram.
Reproduced with permission of the Water Quality Association.

The water treated by reverse osmosis has an improved taste and odor, and several contaminants are reduced by the process, including lead, copper, sodium, nitrates, pesticides, volatile organic chemicals, radium, and arsenic. The level of contaminant reduction will vary, depending on the degree of pollution, the system pressure, and unit maintenance. No system is warranted for complete elimination of all contaminants.

CORROSION CONTROL UNITS

Corrosion is a chemical reaction which affects plumbing fixtures and pipes, causing thinning of the metal surface, staining, and leaching of metals, particularly lead, into the water. The factors that facilitate corrosion are acidity of the water, electrical conductivity, oxygen concentration, and temperature.

Soft water is acidic, with a pH below 7, and conducive to corrosion. Plumbing consisting of two different metals, such as steel and brass, becomes an electrical conductor when the metals are in contact with water containing dissolved minerals, especially hard water. Surface waters contain oxygen that facilitates corrosion, whereas ground waters and deep well water are usually free of dissolved oxygen. The corrosion rate increases when water temperature is increased.

The control of corrosion in household water systems depends on the type of water and other factors involved. Acid waters may be neutralized by special filters containing *calcite* (calcium carbonate) or *magnesia* (magnesium oxide). A solution of *soda ash* (sodium carbonate) added to the water supply by a chemical feed pump is an alternative method for neutralizing acid water.

The plumbing may be protected from the corrosive action of water that is acidic and oxygenated by adding *polyphosphate* compounds or silicates which form a protective lining on the insides of the pipes. Water heaters should be set to avoid temperatures above 140^0F.

DISINFECTION TECHNIQUES

Municipal water supplies in developed countries are treated to control bacterial contamination, and the consumer is generally protected from waterborne infectious diseases. Private water systems are not so well protected, and individual wells may sometimes become contaminated. Periodic bacteriological testing and disinfection may be necessary.

Chlorination. The use of chlorine is the most common method of water disinfection in the United States. Despite its drawbacks, especially the unpleasant taste and odor, clorination is popular in small private water systems when water becomes contaminated. The presence of this chlorine taste and odor is an assurance that disinfection has been completed.

Chlorine solution prepared from household hypochlorite bleach is applied by chemical feed pumps, and is injected into the water line between the well pump and the pressure tank. An activated carbon filter positioned in the water line after the pressure tank will remove excess chlorine and precipitated matter.

If a well becomes contaminated, the source must

be located and removed. Chlorination should be used as a temporary measure and a safety precaution.

Alternatives to Chlorination

Alternative disinfection techniques include ultraviolet light systems, ozone generators, very low levels of silver, and the addition of iodine to the water. All these systems carry disadvantages and some have health risks. The local health department should be consulted when bacterial contamination of the water supply is suspected, and bottled water substituted until disinfection is complete.

National Survey Sponsored by the Water Quality Association. What American restauranteurs are saying about their water supply. Lisle, IL, WQA, 1992.

Ram NM, Calabrese EJ, Christman RF. (Eds). Organic Carcinogens in Drinking Water. Detection, Treatment, and Risk Assessment. New York, John Wiley & Sons, 1986.

United States Environmental Protection Agency. Comparative Health Effects Assessment of Drinking Water Treatment Technologies. Washington, D.C., EPA Office of Drinking Water, 1989.

Idem. Home water treatment units: Fitering fact from fiction. Washington, D.C., EPA, 1992.

Water Quality Association. Point-of-Use Water Quality Answers. Lisle, IL, WQA, Sept 1994.

World Health Organization. Guidelines for Drinking-Water Quality. Vol 2. Health Criteria and Other Supporting Information. Geneva, WHO, 1984.

Idem. Guidelines for Drinking-Water Quality. Vol 1: Recommendations. 2nd Edition. Geneva, WHO, 1993.

GLOSSARY OF TERMS AND ACRONYMS

Aquifier - any subsurface material that holds a large quantity of groundwater and is able to transmit the water readily. There are two types, confined and unconfined.

Bacteria - unicellular organisms having various forms, some of which are pathogens and capable of causing disease.

Carcinogen - a substance known or suspected to cause cancer.

Central Nervous System Effects (CNS) - symptoms of impairment include headache, dizziness, vomiting, blurred vision, seizures, paralyses, weakness.

Cesspool - a covered pit to receive waste or sewage.

Chlorination - a method of disinfecting water by adding chlorine.

Coagulation - formation of a soft or solid mass.

Coliform Bacteria - bacteria that are used as indication of sewage contamination.

Community Water System (CWS) - a system of piped drinking water that has at least 15 connections or serves at least 25 permanent residents.

CWS - community water system.

Contaminant - any substance such as microorganism or chemical present in water causing it to be impure.

EPA - U.S. Environmental Protection Agency.

Epidemiology - study of the incidence, distribution, and control of disease in a population.

FIFRA - Federal Insecticide, Fungicide, and Rodenticide Act, first enacted in 1947 and administered by the EPA since 1970. Under FIFRA, EPA registers pesticide products and ensures that that they will not present unreasonable risks to human health or the environment when used according to label directions.

Ground Water - water found beneath the earth's surface contained in the pores and fractures of soils and geological formations.

Hard Water - water with a high mineral content.

Health Advisory Level (HAL) - the maximum concentration of a contaminant in water that may safely be consumed over a specific time period.

Herbicide - a pesticide used to limit or inhibit plant growth.

Human Health Risk - the probability that a given exposure will damage health.

Inorganic Compounds - compounds which do not contain carbon, e.g. heavy metals and mineral salts.

Insecticide - a pesticide used to control insects.

Leaching - the downward transport through the soil by percolating water of dissolved or suspended minerals, fertilizers, pesticides, and other substances.

Maximum Contaminant Level (MCL) - the maximum permissible level of a contaminant in water that is delivered to any user of a public water system (established by the Safe Drinking Water Act (SDWA)).

MCL - maximum contaminant level.

Micrograms per Liter (mcg/L) - one-millionth of a gram of a substance per liter of water, or parts per billion.

Milligrams per Liter (mg/L) - one-thousandth of a gram of a substance per liter of water, or parts per million.

Nitrate - an oxidized form of nitrogen, plant nutrient and inorganic fertilizer, found in septic systems, animal feed lots, agricultural fertilizers, manured fields, industrial waste waters, sanitary landfills, and garbage dumps.

Nitrite - a form of nitrogen less oxidized than nitrate. An unstable transitional form between nitrate and ammonium.

NPS - National Pesticide Survey.

Office of Drinking Water (ODW) - the EPA office, under the management of the Office of Water, primarily responsible for implementing the Safe Drinking Water Act.

ODW - Office of Drinking Water.

Office of Pesticide Programs (OPP) - the EPA office, under the management of the Office of Pesticides and Toxic Substances, primarily responsible for implementing the Federal Insecticide, Fungicide, and Rodenticide Act.

Organic Compound - natural or synthetic chemical containing carbon.

Parts per Billion (PPB) - one part chemical in one billion parts of water.

Parts per Million (PPM) - one part chemical in one million parts of water.

Pathogens - microorganisms that cause disease.

Pesticide - chemical substance used to destroy,

control, or repel undesirable organisms, including plants, insects, fungi, nematodes, rodents, predators, or microorganisms.

Quality Control - monitoring activities performed by EPA and its contractors to ensure that sampling, data collection, and laboratory procedures are properly conducted and meet specified performance standards.

Raw Water Sample - a water sample gathered prior to treatment of any kind.

Rural Domestic Well - a drinking water well that supplies an occupied private household located in rural areas of the United States, except for wells located on government reservations.

Safe Drinking Water Act (SDWA) - a law passed in 1974 and administered by EPA that establishes national standards for drinking water to provide a safe a wholesome water supply from both surtface and ground-water sources.

SDWA - Safe Drinking Water Act.

Safe Drinking Water Hotline - a toll-free hotline established by EPA in July 1987 to handle requests for information on drinking water issues: 1-800-426-4791 (in Washington, D.C. at (202)-382-5533).

Scientific Advisory Panel (SAP) - a panel of scientists under the authority of FIFRA to advise the EPA on scientific issues related to the assessment of risks posed by pesticides.

Septic System - a sewage system composed of both a septic tank and a septic field. The septic tank is an underground container through which sewage flows very slowly and in which solids separate from the liquid to be decomposed or broken down by bacterial action. The septic field is the area through which the sewage liquid passes and in which it is cleaned through physical filtering by soils, biodegradation, and evaporation.

Soft Water - water with a low mineral content.

Surface Water - water found on the land surface in streams, ponds, marshes, lakes, or other fresh water sources.

Synthetic Organic Compound (SOC) - man-made organic compound, not naturally found in ground water.

SOC - synthetic organic compound.

Total Coliform Count - indicator of fecal pathogens in drinking water.

Total Dissolved Solids (TDS) - all solids dissolved in water, including salts and minerals, measured in mg/L.

Turbidity - cloudiness in untreated or contaminated drinking water caused by suspended sediment. Turbidity may prevent effective disinfection.

Volatile Organic Compound (VOC) - an organic compound that evaporates (volatilizes) readily into the atmosphere and is highly mobile in ground water.

VOC - volatile organic compound.

Water Table - the top of an unconfined (unpressurized) aquifer, below which the spaces in the earthen material are saturated with water.

Well Casing - materials such as concrete, piping, metal, and stone that line and support a well and prevent it from collapsing.